BARN OWL

Barn Owl

DAVID CHANDLER

FIREFLY BOOKS

A FIREFLY BOOK

Published by Firefly Books Ltd. 2011

First printing

Publisher Cataloging-in-Publication Data (U.S.)
David Chandler.
  Barn owl / David Chandler.
[128] p. : col. ill. ; cm.
Includes bibliographical references and index.
Summary: Chapters on subjects such as hunting, courtship and survival, along with personal anecdotes from the author and photographers.
ISBN-13: 978-1-55407-903-2
1. Barn owls. I. Title.
598.97 dc22   QL696.S85C536 2011

Library and Archives Canada Cataloguing in Publication
Chandler, David, 1964-
  Barn owl / David Chandler. -- 1st ed.
Includes bibliographical references and index.
ISBN 978-1-55407-903-2
1. Barn owl. I. Title.
QL696.S85C42 2011   598.9'7   C2011-900545-X

Published in the United States by
Firefly Books (U.S.) Inc.
P.O. Box 1338, Ellicott Station
Buffalo, New York 14205

Published in Canada by
Firefly Books Ltd.
66 Leek Crescent
Richmond Hill, Ontario L4B 1H1

Publisher: Simon Papps
Editor: Elaine Rose
Design: Roger Hammond at Blue Gum Designers
Cover Design: Jason Hopper
Production: Melanie Dowland

Reproduction by PDQ Digital Media Solutions Ltd, UK
Printed and bound in Singapore by Tien Wah Press (Pte) Ltd

Photographs by **AGAMI** (pages 17, 19, 24, 32, 33 (left and right), 34, 39, 43, 51 (top and bottom), 52, 54, 59, 67, 77, 80, 83, 110, 115); **Nigel Blake** (pages 2, 6, 8, 27, 28, 31, 56, 68, 69 (left and right), 72, 73, 74, 75, 112, 116, 123, 124, 128); **David Chandler** (pages 65 (right), 70, 120); **NHPA** (ANT Photolibrary page 23, Stephen Dalton page 93, Michael Leach page 100); **Photolibrary.com** (Hemis page 12, Oxford Scientific pages 15, 20, 103, Age Fotostock pages 48, 94); **Mike Read** (pages 9, 10, 36 (left and right), 37, 41, 50, 61, 65 (left), 78 (top), 84, 85, 86, 88, 97, 98, 104, 106, 107, 108, 109, 111, 117, 119); **Colin Shawyer** (page 76, 78 (bottom)); **Peter Wilkinson** (page 71 (left and right); and **Mike Weedon** (page 47).

Maps on pages 16 and 32 by Stephen Dew. Page 16 based on *Handbook of the Birds of the World*, vol.5. Page 30 based on various sources incluing *Handbook of the Birds of the World*, vol.5 and *The Barn Owl,* by Colin Shawyer.

DEDICATION

This book is dedicated to Chris and Ann Webster,
who have done much for many. And they have
a Barn Owl box on their farm!

ACKNOWLEDGEMENTS

In a work of this kind, the labours of many people who have studied Barn Owls over the years and those who have made their work accessible need to be acknowledged – thank you for providing the facts and figures that are the substance of this book. In this regard, particular thanks are due to Colin Shawyer, Iain Taylor and the late Carl Marti. Thanks to Colin also for helping me with various queries along the way, even pulling over to the side of the road to do so, and for the cache photo.

Peter Wilkinson willingly gave his time and expertise, made valuable connections and was prepared to check the manuscript – thank you Peter. Thanks too to John Lusby at BirdWatch Ireland for his input and for giving me the freedom to publish some of his research findings, and to Nigel Blake, for his Barn Owl anecdotes. Help in various forms came from Dr Alan Poole at the Cornell Lab of Ornithology, Mark Thomas and Kirsi Peck at the RSPB, Jacquie Clark and Paul Stancliffe at the BTO, David Wege at BirdLife International, Kevin O'Hara at the Northumberland Wildlife Trust, Chris and Ann Webster and Ian Llewellyn.

Once again, thank you to Ruth, Amy and Kate, for putting up with another writing project, and in particular to Ruth for her support and understanding.

Finally, thanks to Simon Papps at New Holland for commissioning me again, to Beth Lucas for her help, to Elaine Rose for making the editing process so painless and to Roger Hammond, the designer, who brought words and images together to make another great-looking book.

Contents

Preface

Barn Owls are magnificent birds. They drift effortlessly through fading daylight or punctuate the black of night, stopping birdwatchers and non-birdwatchers in their tracks. They are driver-distracters, with the potential to enliven a tiring motorist more effectively than a double espresso. They are efficient killers that can fly silently thanks to a suite of plumage features and can find their prey in total darkness using just their hearing.

My first encounter with a wild Barn Owl was as a teenager, in 1982. I saw two in one evening, in Suffolk, eastern England, from a car. I don't know how many times I have had the pleasure of seeing them since then but they are still red-letter birds, birds that get underlined in my notebook. It is not just the bird that fascinates – what they leave behind does too. Once, I was working in a school, with group after group of ten and eleven year olds. We were looking at Barn Owl pellets, and one of the children made it clear that in seven years of schooling, this was the most exciting thing he had done.

There is more to the Barn Owl than the fleeting roadside encounters that characterize many experiences of this evocative bird. This book aims to take you deeper into their world. It starts with an overview of the Tytonidae – the family that they are part of – before delving more specifically into the life of the Barn

Opposite: A prolonged view of a Barn Owl such as this is a rare treat, but even a brief encounter with one of these handsome birds is a thrill.

This individual is living up to its name. Barn Owls aren't just found in barns, of course, but they will happily exploit man-made structures to roost or nest in.

Owl, looking at its massive global distribution, the habitats it uses, its movements, what it eats and its design and technique as a hunter. There is information on pellets, territory size, nest and roost sites, breeding biology, life expectancy and more. Its contents are informed by the work of a number of researchers, particularly Colin Shawyer, Iain Taylor and the late Carl Marti. Peter Wilkinson has been working with Barn Owls for many years. I first met him in the mid-1990s, not knowing that I would end up drawing on his Barn Owl experience to write this book. Nigel Blake took many of the photographs that bring the text to life. He has spent many, many hours in the field looking at Barn Owls and some of his observations have found their way into the text too.

I have written this book for anyone in Europe or North America who wants to know more about the Barn Owl. I've tried to make it easy to read but not lightweight – an informative biography. It is not the last word on Barn Owls – they are a well-researched species and there is much more that could be said but, hopefully, whether you have seen one Barn Owl or many, or are still waiting for your first, you will find something between these covers to extend your knowledge and appreciation of this glorious hunter. May it inspire you to get out there and enjoy them for yourself.

David Chandler

Opposite: An airborne Barn Owl is wonderfully buoyant. Rough grassland is good foraging habitat.

1 | The barn owl family

There are over 200 owl species in the world, most of which are in a family of birds called the Strigidae, the so-called 'typical' owls. The Barn Owl (*Tyto alba*), or, to be strictly correct, the Western Barn Owl, is one of a handful of owls that are in a different family, the Tytonidae (the barn owls), and is the only representative of this family in the whole of Europe and North America. Together, the Tytonidae and the Strigidae make up an order of birds known as the Strigiformes, which the taxonomists place between the cuckoos and the nightjars and their allies, not, as you might expect, as close relatives of day-flying hawks and the like. Actually, for more than a century, most scientists did group the owls with the day-flying birds of prey. It wasn't until late in the 19th century that they shifted their position, and since then the luxuriously soft plumage and big eyes of owls, and some other features, have led most taxonomists to conclude that owls are, in fact, close relatives of the nightjars.

Within the barn owl family there are, according to the International Ornithological Commission's (IOC) list of World Bird Names, 18 species. Fifteen of these are *Tyto* owls (*Tyto* is the first part of their scientific name – their genus) and, as such, are very closely related to the Western Barn Owl. The *Tyto* owls can be further subdivided into typical barn owls, masked owls, grass owls and sooty owls (this latter being treated as just one species by the IOC), though, with the possible exception of the Sooty Owl, all bear a strong resemblance to the Western Barn Owl. The other three species are the bay owls and all of these are in the *Phodilus* genus. Our knowledge of the bay owls is poor; indeed, the Congo Bay Owl (*P. prigoginei*) was not discovered until 1951 and it was another 45 years before it was seen again, and it hasn't been seen since.

In this book, the taxonomy and English names used are largely those of the International Ornithological Commission's list of World Bird Names (version 2.4). The exception is the subject of this book! In this chapter, to avoid confusion with other barn owls, or references to the family as a whole, it goes by the name of the Western Barn

Previous page: The heart-shaped face of this inquistive individual is an obvious feature of the barn owl family, Tytonidae.

Owl. Elsewhere in the book, however, it is referred to as the Barn Owl, because, in everyday speech, that is what most people call it.

How are barn owls different from typical owls?
To the average observer of barn owls, the most obvious difference is the shape of the facial disc. In the barn owls, it is heart-shaped, whereas in the typical owls it is round. The facial discs of two of the bay owls (the Oriental Bay Owl, *Phodilus badius*, and the Sri Lanka Bay Owl, *P. assimilis*) are a bit different from *Tyto* owls, however.

Tytonid eyes are large when compared to many other birds, but comparitively smaller than those of the Strigidae, or typical owls, and their beaks are more elongate.

There are other differences but these are not so easy to see. The bills of barn owls are proportionately longer than those of typical owls and their eyes are proportionately smaller. If you are able to examine a bird in the hand or a stuffed specimen, check out its feet. Unlike typical owls, the innermost toes of barn owls are more or less the same length as their middle toes and their middle claws have a comb-like structure on the side facing the bird. This is a preening tool and is meant to be a distinctive barn owl feature, but there are records of individuals of more than one barn owl species that lack it and in Western Barn Owls this feature takes time to develop.

There are other differences between typical owls and barn owls. The wishbone of a barn owl is joined to its breastbone but on typical owls it isn't – but that's a difference that you won't spot when you are out birdwatching!

Distribution of all species in the barn owl/grass owl/ masked owl family.

Where are barn owls found?

As a family, barn owls make a living on every continent bar Antarctica. They are even found on some remote islands. You won't find any in the far north, much of Asia, parts of the Arabian Peninsula, the Sahara and New Zealand.

With ten species, the Australasian region is the most barn owl diverse part of the world. Australia itself has four species: the Western Barn Owl, the Australian Masked Owl (*T. novaehollandiae*), the Sooty Owl (*T. tenebricosa*) and the Eastern Grass Owl (*T. longimembris*). When it comes to barn owl diversity, North America and Europe, with just one species, are impoverished regions.

A lowland encounter with a Western Barn Owl in Europe or North America is not entirely representative of the family as a whole. At its heart, the Tytonidae are birds of the tropics, and the majority of its members are birds of tropical forests and neighbouring areas. Indeed, in some areas, Western Barn Owls are rainforest birds. Clearly, though, the family isn't limited to the tropics and reaches much further north and south too – from Canada, Scotland and Denmark in the north to Tasmania, the Falkland Islands and almost the southernmost

An Australian Masked Owl, the longest and heaviest member of the barn owl family.

tip of South America in the southern hemisphere. Neither is the family restricted to the lowlands – the Sooty Owl is found as high as 4,000m (13,000ft) above sea level in the mountains of New Guinea, and, perhaps surprisingly, in the Andes, the Western Barn Owl can be found at similar altitudes. The Congo Bay Owl (*P. prigoginei*) is poorly known but has been seen at over 2,400m (7,800ft).

Both of the grass owl species, the African Grass Owl (*T. capensis*) and Asia's and Australasia's Eastern Grass Owl (*T. longimembris*) also live quite

contentedly at high altitudes, peaking at 3,200m (10,500ft) and about 2,500m (8,200ft) respectively. These species are found in more open habitats, including grasslands and marshes.

What do they look like?

This is not the most diverse of the world's bird families. Most of the species look like variations on the Western Barn Owl, with underparts that have varying amounts of dark spotting on a white to orangey-buff background, facial discs that, in whole or in part, may be white, off-white, brown or orangey-buff, and upperparts painted from a palate of greys, browns, black and orangey-brown, with some white spotting. Typically, the males are smaller and paler than the females, though a female's colours fade as she gets older. The females are also more obviously spotted, both above and below.

The species that depart from the Western Barn Owl template the most are the bay owls, an example of which can be seen opposite. The Oriental Bay Owl and the Sri Lanka Bay Owl look the most different from the Western Barn Owl and they are, of course, in a different genus – *Phodilus*. The Congo Bay Owl is in that genus too but looks much more like one of the *Tyto* owls than the other bay owls do.

In terms of length, the smallest members of the family are the bay owls, small individuals of which are just 23cm (9.1in) from bill tip to tail tip. Note, however, that there is no description of the male Congo Bay Owl, so, assuming this species follows the family trend described above, he could be even smaller. Female Congo Bay Owls weigh 195g (6.9oz) and the males might be even lighter.

Currently, even though it is longer, the recorded lightweight in the family is the Western Barn Owl, which, at its smallest, tips the scales at a surprisingly light 187g (6.6oz). The barn owl heavyweight is the Australian Masked Owl. There is lots of variety within this species, in both size and colour, and a very pronounced size difference between the male and the female. A small male weighs 420g (14.8oz) but, at 1,260g

(44.4oz), a big female is three times heavier – the record breakers are found among the Tasmanian subspecies. These birds are the longest in the barn owl family too, reaching a length of 57cm (22.4in).

An Oriental Bay Owl. The bay owls are in a different genus to most of the barn owl family.

Definitely not vegetarians

Our knowledge of the barn owl family is patchy and that includes our understanding of what they eat. Lots of work has been done on the diet of the Western Barn Owl and this is explored in some depth in chapter four. At the other end of the scale, however, no information at all is available on the culinary habits of the Minahassa Masked Owl (*T. inexspectata*), the Taliabu Masked Owl (*T. nigrobrunnea*) and the Moluccan Masked Owl (*T. sororcula*), all of which have very limited distributions on Indonesian islands, or of the Manus Masked Owl (*T. manusi*), which lives on Manus Island off Papua New Guinea, or of Africa's Congo Bay Owl.

We know a little about what the Golden Masked Owl (*T. aurantia*) eats. This species lives on New Britain, another island off Papua New Guinea, and, on dissection, one was found to have eaten some kind of rodent. This is not a great surprise – the barn owls' beaks and talons are a clear indication of a carnivorous diet and, as far as we know, mammals are what the birds in this family eat most of. This includes shrews and rodents, which are particular favourites of the grass owls and typical barn owls. The Sooty Owl has a diet that includes mice, bats, wallabies, antechinuses (shrew-like marsupials), dunnarts (another group of small marsupials), pygmy possums and ringtail possums, the latter of which can tip the scales at 900g (31.7oz) – nearly 80 per cent of the weight of the biggest Sooty Owl. This is a bird of forest interiors but they do sometimes hunt in more open habitats. Like the Sooty Owl, the Australian Masked Owl is an enthusiastic marsupial-muncher!

In Australia, rodent plague years provide an abundance of food for barn owls, which is exploited en masse by an assortment of Eastern Grass Owls, Western Barn Owls and Australian Masked Owls.

Hunting techniques in the family include 'perch-hunting', where prey is detected from a perched position, and low, quartering flights, from which the hungry owl will drop on to its prey, both of which are employed by the Western Barn Owl, the Australian Masked Owl and the Ashy-faced Owl (*T. glaucops*). The Sooty Owl grabs its prey in the treetops as well as on the ground and does not need any light to find its

Previous page: This North American Barn Owl, photographed in Arizona, is likely to feed on a diet of local rodents. Pocket mice are important prey in this part of the United States.

food. The Oriental and Sri Lanka Bay Owls are also predators of the treetops, and their feeding techniques include launching themselves from tree trunks to snatch a meal in mid-air. No one knows what the Congo Bay Owl does. While the presumption is that barn owls are primarily mammal-eaters, they do take other food too, including birds, reptiles, amphibians, fish, insects, spiders and scorpions. The Australian Masked Owl is known to have eaten kookaburras; the Ashy-faced Owl has fed on hummingbirds and trogons; and the African Grass Owl has swallowed eggshells. Carrion is sometimes consumed too.

Raising a family

Predictably, there are lots of gaps in our knowledge of barn owl breeding biology. From what we do know, barn owls are territorial and stay in their territories all year. The Australian Masked Owl and the Sooty Owl are particularly territorial. Whatever the time of year, somewhere, barn owls will be breeding. The timing of breeding varies from place to place, but Western Barn Owls, for example, given a global overview, can be breeding in any month, though those in North America and Europe tend to breed between March and June, or into August if they have a second brood, though some second broods are even later in the year with fledging taking place in October or November. In the interior of Australia, the Eastern Grass Owl doesn't have a fixed breeding season but seizes the moment as conditions allow. On the coast it breeds from February to September; further north, in New Guinea, its season is contracted to May and June; and in Asia breeding is mostly between October and March. The other grass owl, the African Grass Owl, times its breeding so that it can make use of decent grass cover. This is

A Sooty Owl, one of the more territorial members of the family.

available at different times in different parts of its range and, as a consequence, somewhere in Africa, whatever the month, African Grass Owls will be breeding. The main Sooty Owl breeding season stretches from January to September.

An amorous male Western Barn Owl becomes more vocal and may indulge in display flights as part of his courtship behaviour (Australian Masked Owls are thought to use similar ploys). He brings food to the female to fatten her up for egg-laying and to strengthen the pair bond. At the nest site the pair preen each other, 'fence' with their bills and utter a strange assortment of noises. A more detailed account is given in chapter five.

Typically, barn owls are ground-nesters or cavity-nesters. As their name suggests, both of the grass owls nest on the ground, and their nests are similar. African Grass Owls frequently select nest sites close to or beneath bushes. Their nest is a mat of flattened grass, accessed via a tunnel made by the owls, which runs through the grass. Additional tunnels allow the owlets and their parents to flee from hungry predators.

From what we know of the rest of the barn owl family, the other species nest mostly in cavities of one sort or another, though data is non-existent or nearly so for at least half of the species. Holes in trees are frequently used as nest sites, ranging from just 2m (6.6ft) up, in some Oriental Bay Owls for example, to a lofty 50m (about 160ft) up for some Sooty Owls. It is not just tree-holes that are used. Caves, cliffs and sinkholes have also been used as barn owl nest sites, as well as a host of unnatural settings, including castles, churches, bridges and various derelict buildings. The Western Barn Owl nests in a wide variety of natural and unnatural places, including nest boxes – more information is included in chapter five.

No real effort is made to construct a nest, though the eggs of Sooty Owls and Western Barn Owls may rest on a bed of regurgitated pellets and other nest site debris. Western Barn Owls are known to make a shallow scrape for the eggs in a bed of shredded pellets. Sooty Owls sometimes nest in caves and many use the same site year after year. When

Opposite: African Grass Owl chicks. This ground-nesting species does not have a well-defined breeding season, but breeds whenever conditions are favourable.

this happens, food remains pile up on the floor and give ornithologists a means of estimating how long the cave has been in use – the Jenolan Caves in New South Wales, Australia, have accommodated breeding Sooty Owls for thousands of years.

Usually, a freshly laid barn owl egg is white and clutch sizes range from 1 to 16 eggs. The smallest clutches that we know of are those of the Sooty Owl and the Australian Masked Owl, both of which usually lay two eggs in a clutch. The biggest clutches belong to the Western Barn Owl, which lays clutches of between 2 and 16 eggs, though 4–7 is normal and exceptionally large clutches may have been produced by two females. Based on the Western Barn Owl, scientists suspect that females do the incubating, which can last from 29 to 42 days. Once they are out of the egg it can be anything from about 42 to about 90 days before the owlets fledge and it could then be as much as another 6 months before the young owls can fend for themselves fully.

Not adventurous travellers

Generally speaking, barn owls are sedentary birds that stick to their territories, though they may trespass on someone else's patch outside of the breeding season if prey is hard to come by. After breeding some barn owls disperse, and some fairly impressive long-distance movements have been recorded, but ornithologists think that most birds don't go very far. Having said that, an African Grass Owl showed up in Ethiopia, 900km (about 560 miles) from where you might expect to see one, and a ringed Western Barn Owl moved 1,000km (621 miles) from Senegal to Sierra Leone. Most dispersing barn owls are thought to be in their first year of life.

Barn owls that live in Australia do things a bit differently, with some Eastern Grass Owls and Western Barn Owls adopting a nomadic lifestyle to enjoy the easy pickings of cyclical rodent plagues. Some individuals that spend the winter around the coast move inland in the summer and there are eastward and northward movements in the winter too. Some Australian Masked Owls are also nomadic.

Opposite: Wherever this White-breasted Barn Owl is going, it's not likely to be too far from home. Most members of the barn owl family are quite sedentary, only rarely dispersing over long distances. In Australia, however, some barn owls are nomadic and take advantage of the easy food supplied by rodent plagues.

2 | Introducing the Western Barn Owl

From here on in this book, the Western Barn Owl is referred to simply as the Barn Owl. As you can see from the map below, it has an impressive global distribution. Its range includes most of the Americas, Caribbean Islands, much of Europe, parts of North Africa, most of sub-Saharan Africa, parts of the Arabian Peninsula, parts of Asia, Australia, Tasmania and various islands in the Atlantic, Indian and Pacific Oceans, including the Galapagos, the Falklands, the Cape Verde Islands and Madagascar. It is, in fact, among the world's most widely distributed land birds, if not the most widely distributed.

The Barn Owl's scientific name is *Tyto alba*. *Tyto* comes from *tuto*, which is the Greek word for owl, and *alba* is Latin for white. Literally then, its scientific name means 'white owl'. Taxonomists recognize many

Distribution of the most widespread Barn Owl subspecies: *alba* (red); *guttata* (purple); *erlangeri* (orange); *affinis* (turquoise); *pratincola* (yellow); *tuidara* (blue); *javanica* (brown); *stertens* (lilac); and *delicatula* (green); range for all other subspecies is shaded grey.

barn owl subspecies or races, and different authorities come up with different totals. The Clements checklist of April 2010 (one of about four published lists of the world's bird species) lists 30 subspecies, though two of these are treated as separate species on the IOC list. There is no single, definitive list of species and subspecies, and some of today's subspecies may well be promoted to tomorrow's species. Each subspecies has a suffix added to *Tyto alba*. Some have tiny distributions and are limited to small islands. *Poensis*, for example, is included as a subspecies in the Clements list and is found only on Bioko, an island in the Gulf of Guinea that measures about 70km (43 miles) by 32km (20 miles). Others have much larger ranges – this includes affinis, which is the only race found on mainland Africa south of the Sahara.

This book is concerned primarily with Barn Owls in Europe and North America. In these parts of the world, there are, according to the Clements list, seven subspecies. Four of these are limited to island groups in the Mediterranean or Atlantic and will not be given much attention. Two subspecies are found in mainland Europe and the UK. These are *alba* and *guttata*. One is found in North America – this is *pratincola*. These three subspecies are the main focus of this book.

Where are they found?

Pratincola is Latin for meadow inhabitant. Its range stretches from Canada through the mainland United States into northern Mexico. According to the Clements list you can find them on the Bahamas, Bermuda and Hispaniola too. For ease of reading, *pratincola* is referred to here as the North American Barn Owl.

In mainland Europe, *guttata* is found north and east of a line that runs through Belgium, eastern France, western Switzerland, northern Italy, the northern end of the Balkan countries, northern Bulgaria and northern Turkey. *Alba* is found to the south and west of this line. The line approximates to the 3°C January isotherm, and around this line Barn

A White-breasted Barn Owl, the *alba* subspecies, found south and west of the 3°C January isotherm.

Owls of the two different forms interbreed and show characters somewhere between pure *alba* and pure *guttata*. *Alba* can also be seen on Sicily and the Balearic Islands. The map on page 30 shows the distribution of these two subspecies. Again for ease of reading, in this book *alba* is the White-breasted Barn Owl and *guttata* the Dark-breasted Barn Owl.

Do they look very different?

Text-book White-breasted and Dark-breasted Barn Owls do look different. Essentially, the latter have obviously greyer upperparts and buff to orangey underparts, with lots of small, dark spots (*guttata* means spotted). Compare this to the basically white underparts and paler upperparts of white-breasted birds. The facial disc of a Dark-breasted Barn Owl is not as white as a White-breasted Barn Owl's and has noticeable orangey tones, particularly around the eyes. Also, the stiff feathers that frame the facial disc – the ruff – are more conspicuous. On the lower wing surface, look for the buff to orangey underwing coverts of Dark-breasted Barn Owls. These are white on White-breasted Barn Owls.

The pictures on pages 49 and 53, however, show that things are not always this clear-cut.

Note, too, that White-breasted Barn Owls do have some orangey tones near their eyes and these may even surround the eyes, and it is not unusual to see individuals with an orangey-buff wash at the edge of the breast. White-breasted Barn Owls can be a lot darker than many people realize. It is possible that the UK white-breasted population has been influenced by the genes of dark-breasted birds released from captivity – so there may be birds which show characters somewhere between the two races.

North American Barn Owls are very variable in appearance, with upperparts a mix of orangey-buff, shades of grey, brown and white, with plenty of black and white spots. Underneath, their colour can be anything from white to orangey-buff. Some have no spots on their underparts; others have lots of dark spots, and the spotting can extend on to the underwing coverts. The facial disc is white, though it may have obvious burgundy tones, particularly near or around the eyes and down towards the bill. This subspecies can be significantly bigger than White-breasted and Dark-breasted Barn Owls, a big female being 46 per cent heavier than a big Dark-breasted Barn Owl and approaching three times the weight of a small White-breasted Barn Owl. North American Barn Owls can be significantly longer too

Dark-breasted Barn Owl, *guttata* (on the left) and White-breasted Barn Owl, *alba* (on the right).

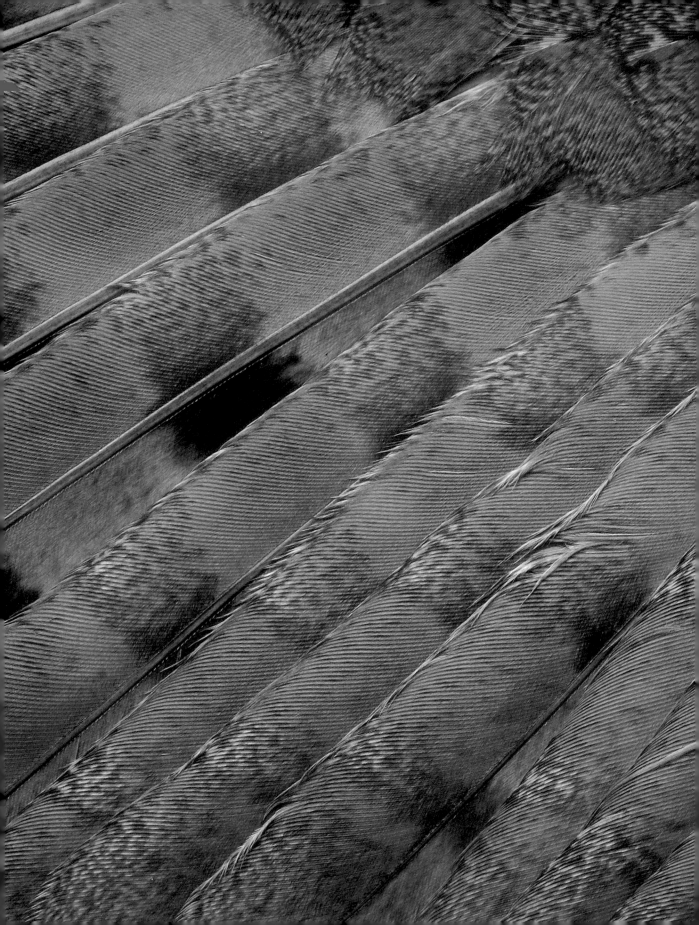

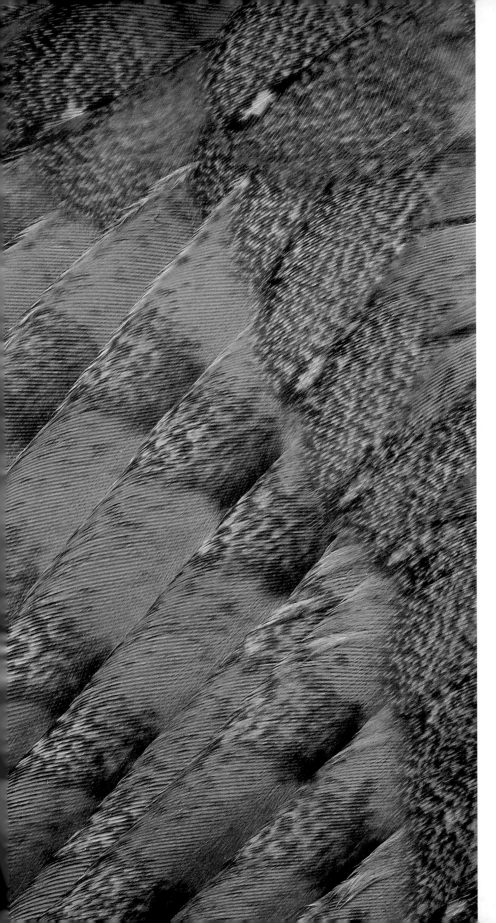

Wing feathers of a Dark-breasted Barn Owl. This subspecies, found north and east of the 3°C January isotherm, typically has greyer upperparts than the White-breasted Barn Owl. In the UK, some individuals may show intermediate characters, possibly because White-breasted Barn Owls have bred with Dark-breasted birds released from captivity.

and have stouter feet and legs. The table below summarizes the size differences. It pulls together data from several sources but it is not clear whether the textbook length for Barn Owls in Europe refers to one or both of the subspecies.

Male or female?

Generally speaking, the males are paler than the females and the females are bulkier with more speckling on their underparts, but working out the gender of a bird in the field is not easy.

A comparison of the sizes of subspecies of the Western Barn Owl

Subspecies	Male weight	Female weight	Length
White-breasted Barn Owl	240–360g (8.5–12.7oz)	245–435g (8.6–15.3oz)	33–35cm (13.0–13.8in)
Dark-breasted Barn Owl	250–400g (8.8–14.1oz)	252–480g (8.9–16.9oz)	33–35cm (13.0–13.8in)
North American Barn Owl	400–560g (14.1–19.8oz)	420–700g (14.8–24.7oz)	32–40cm (12.6–15.7in)

In the UK, Colin Shawyer concluded that when a pair of White-breasted Barn Owls is seen together, the female looks bulkier and, almost without exception, has more flecking on her underparts and darker upperparts. An obvious white 'collar' on each side of the neck also helps to identify the male. Birds seen in isolation are trickier. If the underparts can be seen well enough, the amount of flecking can be a very helpful clue to a bird's gender. Iain Taylor's work in Scotland found flecks on 98 per cent of female White-breasted Barn Owls, including their underwing coverts. Only 5 per cent of males had any underpart flecking. The number and size of the flecks is variable. Female White-breasted Barn Owls range from those with many big, lozenge-shaped flecks all over their underparts to those with just a handful of small marks on the flanks. Males in the 'flecked 5 per cent' did not have many flecks and those they had were just thin marks. The underwing coverts are a very good pointer – they are normally flecked on female White-breasted Barn Owls but

Taylor found only 2 per cent of males had any underwing flecks. Note that the 'flecking technique' may not work well with recently fledged White-breasted Barn Owls.

Male Dark-breasted Barn Owls tend to have paler orange tones on their underparts than the females and, like White-breasted Barn Owls, less flecking. North American Barn Owls can usually be sexed by the colour of their breast and facial disc, the pale birds being male. Individual variation, however, means that it is not always easy to work out whether you are looking at a male or a female. A dark male can look like a pale female, though the palest males have unblemished, pure white underparts and dark females have plenty of obvious flecks on richly coloured buff underparts. The amount and size of breast flecking provides clues – the females normally have more and bigger flecking than the males – but is

Glowing in the dark

Imagine encountering a Barn Owl that literally glowed in the dark! That's what happened to tens of witnesses in Norfolk in the east of England early in the 20th century. Here are some extracts from an original account in the transactions of the Norfolk and Norwich Naturalist's Society:

… on February 3, 1907, on reaching the top of Twyford Hill, we noticed a light apparently moving in the direction of Wood Norton… After moving horizontally backwards and forwards several hundred yards, it rose in the air to the height of forty feet or more… The light was slightly reddish in the centre, and resembled a carriage lamp, for which we at first mistook it. We watched it for twenty minutes and were quite at a loss to ascertain its cause. The light emerged across the field, at times approaching within fifty yards of where I was standing. It then alighted on the ground for a few seconds… On another occasion, the evening being dark, the bird issued from a covert. Its luminosity seemed to have increased and it literally lighted up the branches of the trees as it flew past them…

There were other records too, but none for many years, and all of the reports are of Barn Owls in Norfolk. Various explanations have been offered, including tiny bits of luminous Honey Fungus *Armillaria mellea* or luminous bacteria, both of which could be found in old trees, sticking to the birds' plumage.

This Barn Owl, photographed in Belgium, shows features between pure *alba* and pure *guttata*. There is obvious flecking and a conspicuous ruff – *guttata* features. The underwing coverts are largely white, however, and the underparts of its body are paler than on a typical *guttata* bird.

not as good a guide as the bird's colour. Note, too, that males of this subspecies have some, and sometimes many, flecks on their underwings and perhaps buff tones too, so Taylor's technique does not work so well west of the Atlantic. In the North American subspecies, the burgundy tones in the facial disc also tend to be more intense in the females.

Young or old?

To the average observer, once a Barn Owl has fledged and left the nest it looks the same as the adults. Colin Shawyer observed a buff wash on the breast, or at the sides of the breast, on some of his study birds in the UK. This was easiest to see on juvenile females, but once they had moulted it was harder to see.

Birds in the hand can be aged. This is possible because of the way they moult their flight feathers. Once a bird has left the nest, it doesn't moult until next year's breeding season. At this first moult they normally only replace one flight feather – primary six (the primaries are the big feathers on the part of the wing nearest the tip), though primary seven may be moulted too. Four to six more primaries are usually moulted a year later, specifically those immediately left and right of primary six. At their next moult, which is now three years after hatching, the rest of the primaries are replaced. To spot the new feathers the wing needs to be spread and the underside checked. The newer feathers are whiter and shinier than the older ones. The table below summarizes the typical pattern.

New for old – moult

A Barn Owl's life takes its toll on its plumage and, from time to time, feathers need to be replaced. Its moult strategy is not a straightforward annual replacement of all of its feathers. This account may not prove to be the last word on Barn Owl moult but hopefully provides a reasonably accurate, generalized overview.

The appearance of primary feathers according to age

Year	Primaries
First. Hatching to first breeding season.	All look the same.
Second. First breeding season to second breeding season.	Primary six looks new (and sometimes primary seven).
Third. Second breeding season to third breeding season.	Several primaries look new.
Fourth. Third breeding season to fourth breeding season.	Several primaries look new.

Opposite: A preening White-breasted Barn Owl. Ultimately, worn and damaged feathers are replaced by moult, but in the meantime, feather care is vital. Barn Owls have ten obvious primaries on each wing and one that is not obvious. This eleventh primary is tucked away under the wing coverts at the base of the primary next to the outermost one. In first-year birds this is white too.

It takes years to replace all of the flight feathers (the primaries and secondaries). In its second calendar year, during its first moult, a Barn Owl replaces just one or two primaries and up to four secondaries. Moult number two sees another 4–6 new primaries. Most of the remaining old secondaries are also moulted. During the next moult, in the owl's fourth calendar year, any primaries that still haven't been replaced get their turn, as may some old secondaries, and some birds begin the cycle again. Moult in subsequent years, for any birds that survive this long, could be even more complex!

New tail feathers are produced during the second calendar year but not all of them are replaced in one moult. There are 12 tail feathers, and the first to go are the middle two and the outer two. Some tail feathers are replaced during later moults, though, so far, no one has worked out any predictable process in this part of the tail moult. Moulting every tail feather in one season has been known and so has moulting absolutely none of them.

Most of a bird's feathers, however, are not the flight and tail feathers but the much smaller, softer feathers that cover the head and body, and the wing coverts. One study in Switzerland found that at least some, and sometimes all, underpart body feathers were replaced during an annual cycle. Taylor's work in Scotland, however, concluded that body feathers were moulted every two or three years.

Typically, Barn Owls moult between May and October (November in North America). The females begin before the males, taking advantage of the time that they are on eggs and looking after the young, and starting the process when they are at their heaviest (though particularly early breeders defer moulting until later in the breeding season). For the male this is a very busy time, with a partner and, later, some owlets depending on him for food. Moulting and growing new feathers takes energy, and the male sensibly hangs on to his feathers until his offspring are at their heaviest and the need to bring in the meat has diminished.

These Barn Owls were photographed in Belgium. Like the bird on page 39, their features are intermediate between pure *guttata* and pure *alba*.

Moult is different for tropical Barn Owls, at least for those in Malaysia. These start the process, on average, at about ten months of age and moult their entire collection of flight and tail feathers in roughly seven months. After this, moult is a once-a-year job.

Screeching and snoring – Barn Owl noises

The Barn Owl's vocal repertoire is diverse, including twitters, chirrups, hisses, squeaks, clicks and, perhaps best known of all, some impressive screeches and snores. One thing they do not do is hoot. North American Barn Owls seem to be pretty similar to European Barn Owls when it

Featherprints

The patterns on a Barn Owl's flight and tail feathers are different on every bird. When they moult, the feathers are left behind and this can help Barn Owl workers find out if a bird is still using its favourite nesting or roosting site.

comes to the noises they make, though most of the work in this area has been done in Europe.

It is not without reason that the Barn Owl has also been known as the 'Screech Owl'. Sometimes they screech just once, but at other times they screech again and again. They screech as part of their courtship behaviour – they do it perched and they do it in flight, and both male and female are screechers. Courting birds engage in courtship flights near a potential nest site. They screech as they fly, particularly the male, whose call is crisper, slightly shorter and more soprano than the female's. The screech gets louder, rattling or quivering, and suddenly stops. The female's rendition is more quavering and doesn't flow as well as the males, getting quieter as it finishes. It is a far-carrying call that, in European birds, is about two seconds long. In contrast, work in Ohio by McClean and Colvin found that Barn Owls screeched for less than half a second, on average. A male may also give a shrill scream as he approaches the nest, to alert his partner and offspring to danger.

The 'Screech Owl' snores too, especially the owlets and females. The snore is rough and wheezy and, like the worst human snorer, goes on and on, though each individual snore lasts no more than about a second. It translates into 'I'm here' and 'go and get some food', when the female is 'talking' to the male, and 'come back for some food' and, later, 'come on out', when she is addressing her offspring. The nestlings use it to say 'I'm here' and 'get me some food'. The hungrier the nestlings are, the worse the snoring, and when a food-bearing parent turns up, the nest is a cacophony of frantic snoring!

Repeated hisses are used to warn off would-be predators. These last three or four seconds and are sometimes reinforced with loud clicking noises. The nearer the predator gets, the more frenzied the hissing becomes. In North America it is mainly the owlets that hiss in self-defence, whereas in Europe the female is a much more active participant. Despite their impressive vocal repertoire, when they are not pre-occupied with breeding, which equates to about six months of the year, Barn Owls don't make a lot of noise.

What is their conservation status?

BirdLife International estimates the world's population of Barn Owls as five million birds (this includes two subspecies that the IOC treat as species, the Eastern Barn Owl, *T. javanica*, and the Andaman Masked Owl, *T. deroepstorffi*). They have a big population and a very big range, and their numbers seem to be holding their own. All of which means that, at a global level, the conservation scientists are not worried about Barn Owls – they are classified as of 'least concern'.

Its status in the United States ranges from good to endangered. Overall, it is a common species, but its status varies from place to place. It is doing well, or reasonably so, in some areas but not everywhere. Barn Owls in the Midwest and parts of the east have not fared well, and this wonderful hunter has found its way on to the endangered species list in a number of states, including Wisconsin, Iowa, Missouri, Illinois, Ohio, Michigan and Connecticut, and is a 'species of special concern' in some others, such as the Dakotas, Nebraska, Minnesota, New York, New Jersey and Massachusetts.

Between 110,000 and 220,000 pairs are thought to breed in Europe. This sounds good, but numbers fell from 1970 to 1990 and its fortunes varied over the next decade. It did well in some countries, with increases in Belgium, the Netherlands, Denmark and Germany for example, but badly in others, with declines in Poland, Italy, Spain, the UK and Ireland, among others. Overall, its numbers are thought to have dropped during this ten-year period too. Because of this ongoing loss, BirdLife International has evaluated the Barn Owl as of 'European Conservation Concern'. It is a 'SPEC 3' bird – a 'Species of European Conservation Concern, category 3'. Apart from 'Non-SPEC' species, which are not of conservation concern, this is the least-worrying category; birds that are believed to be at greater risk are 'SPEC 1 or 2'.

In the UK, Birds of Conservation Concern 3 (2009) gives the Barn Owl amber status, one down from red. The only reason it has this status, however, is because of its SPEC status at a European level. The most recent survey of British Barn Owls took place between 1995 and 1997 and concluded that there were between 3,000 and 5,000 pairs. Barn Owl workers believe this to be a conservative figure and estimate the 2010 population at more like 7,000–8,000 pairs.

Mobbing

A Barn Owl that is out and about during daylight is not always greeted with approval by other bird species. Unlike the Tawny Owl (*Strix aluco*) and the Barred Owl (*S. varia*), it is not usually mobbed by small birds, though Nigel Blake has witnessed Eurasian Blue Tits (*Cyanistes caeruleus*) mob Barn Owls very determinedly, especially when the owl settled under their nest box. Members of the crow family, including Northern Ravens (*Corvus corax*), Carrion Crows (*C. corone*), Rooks (*C. frugilegus*) and Eurasian Magpies (*Pica pica*), day-flying birds of prey, gulls and other species will give Barn Owls a hard time, with airborne birds getting most of the attention. A picked-on Barn Owl does not respond aggressively. It might just perch somewhere or head for cover, either of which seems to satisfy its pursuers.

Kestrels and Barn Owls

In Europe, Barn Owls and Common Kestrels (*Falco tinnunculus*) can be present in the same area, hunting the same prey. They sometimes nest in close proximity with no apparent problems, but conflicts over food are certainly not unknown, with both species trying to steal food from the other.

There are some spectacular tangles, with, for example, the falcon coming in under the owl, from the head or tail end, and flipping over so that its feet are uppermost to snatch what would be the owl's next meal. Fights take place on the ground too – a Common Kestrel has been known to target a Barn Owl that was mantling its prey (hiding it under spread wings). There are a few reports of mortal combat, with three records of Common Kestrels being despatched by Barn Owls.

A mid-air tussle for food. Barn Owls and Common Kestrels may occupy the same area which sometimes results in a spectacular competition for prey.

3 | Home and away: habitat and movements

Good Barn Owl habitat needs to provide plenty of food and suitable nesting and roosting sites. Essentially, the Barn Owl is a bird of open habitats, including farmland, grasslands, marshes and deserts. Food comes in the form of small mammals, which, typically, it hunts in rough grassland or other rank vegetation. In the UK, for example, unmanaged grassland at the edge of fields and woods, on woodland tracks, and at the side of ditches, hedges and roads provides good habitat for voles and shrews, and can therefore be a great hunting area for a hungry Barn Owl. John Lusby's work on Barn Owls in Ireland has found that cereal crops can be an important foraging habitat but, at the time of writing, he is not sure why. It does not seem to be connected with farming intensity, as some of the areas are farmed fairly intensively and others less so.

Man's activities have not always been a blessing to Barn Owls, but buildings and, more recently, purpose-built Barn Owl boxes have provided a very good alternative to their more natural roosting and nesting sites in holes in trees, cliffs and banks. This is not a bird that is averse to living alongside people and sometimes lives in very close proximity – when buildings on working farms are used to roost or nest in, for example. Some Barn Owls live in cities and they have even nested in the New York Yankees' stadium.

In Europe, Barn Owls are not usually found at more than about 600m (1,970ft) above sea level, though a French pair went against the norm and bred at an altitude of 1,500m (4,920ft). British and Irish birds are very much birds of the lowlands. In the 1980s Colin Shawyer looked at data from 2,699 nests. Ninety-two per cent of them were less than 150m

Previous page: Woodland tracks and areas close to the edge of woods can be a fruitful source of the small mammals that make up the bulk of the Barn Owl's diet.

Nesting boxes provided for Barn Owls in Lincolnshire, England.

Above: Barn Owls often frequent farmland, especially sites around the edges of fields, which can be a plentiful source of food.

Left: This species is comfortable around areas of human habitation and may make its home in farm buildings and outhouses as well as other man-made constructions.

(about 490ft) above sea level. He has also found that, in Britain at least, Barn Owls often live fairly close to waterways, with many nests no more than about a kilometre (0.6 mile) from a stream or river. When food is harder to come by, Barn Owls will move into different areas and can even be found more than 800m (2,600ft) above sea level. They are found at higher altitudes in North America, with 1,800m (5,900ft) being the limit for birds in Colorado, but in local terms this is not particularly high. This is nothing, of course, compared to the Barn Owls who make their living 4,000m (13,000ft) up in the Andes.

Do Barn Owls migrate?

Overall, European and North American Barn Owls are very sedentary birds. In Europe, even Barn Owls that breed at high latitudes stay put in the winter, which can prove fatal in particularly cold years. It has been

These two individuals were photographed in the Netherlands. Their Dutch name, 'Kerkuil', translates as 'Church Owl', reflecting their frequent use of these buildings on mainland Europe.

said that some birds in northern North America migrate, with supposed migrants passing through Cape May in New Jersey and, previously, through Ohio. Birds are known to have flown 2,000km (1,250 miles) south. The return flight north has not been confirmed, however; the only evidence for it is circumstantial – no ringed (banded) birds have been found in the summer over 320km (200 miles) south of where they hatched. It could be that the birds that head south are not migrants but young birds dispersing after the breeding season.

Ringing recoveries suggest that very few British Barn Owls leave the country and few Continental birds find their way to Britain, though there are records of Barn Owls from the Netherlands, Belgium and Germany that have survived the North Sea crossing and single British Barn Owls have turned up in France, Germany and Ireland.

Dispersal

Young Barn Owls do not leave the parental home immediately and may stay in the area for up to about two months. When they do disperse, the youngsters could go in any direction, though local geography, be it uplands or large areas of water, seems to affect their direction and how far they travel. It is not unusual for siblings to head off on different bearings from each other, though some disperse in the same direction. In mainland Europe there is a tendency for dispersing Barn Owls to head to the south and west but British Barn Owls disperse in all directions. This is also true of North American Barn Owls, though, as mentioned above, there is a southerly bias to the movements of birds from some northern populations.

Dispersal distances vary. One of two siblings from Ohio travelled 1,070km (about 670 miles) north-east, while the other went 800km (500 miles) south-east. Birds from northern states seem to disperse further than those from southern states, most of which are found no more than 80km (50 miles) from where they hatched. More than 40 per cent of northern birds move further than this, and from records of birds that are at least six months old, 27.7 per cent had moved over 320km (200 miles). Carl

Marti's work in Utah found that females dispersed further than males, settling to breed an average of 61.4km (about 38 miles) from where they started life. The males settled 35.7km (about 22 miles) away. The record for a Utah nestling was a dispersal distance of 1,267km (about 790 miles). Iain Taylor's work on Scottish Barn Owls and Roulin's work in Switzerland also found that the females dispersed further.

British Barn Owls do not disperse very far. Ringing (banding) data for Britain and Ireland gives an average dispersal distance of only 12km (7.5 miles). Around 90 per cent of British Barn Owls find a breeding site no more than 50km (about 31 miles) from their hatching place, and 53 per cent settle no further than 10km (about 6 miles) away. Some Barn Owls on the European mainland are more adventurous. The 50km (about 31 miles) figure for Denmark is similar to Britain but drops to 70 per cent in the Netherlands and France and to less than 40 per cent in Germany, where over 30 per cent had moved more than 100km (about 62 miles).

Ringing (banding) a young chick. Ringing data provides a wealth of information about dispersal habits. The vast majority of Barn Owls in Britain establish a breeding site within 50km of where they hatched. This figure varies more widely for the Barn Owls of mainland Europe, which occasionally travel great distances.

In Britain, dispersal distances vary little from year to year, but this is not the case in Continental Europe. Here, some years see more birds dispersing over greater distances. These years are known as…

The wanderjahre

Wanderjahre are thought to be triggered by a rapid drop in vole numbers after a bumper breeding season, a combination that means there is not enough food to go around. Young birds are forced to move further than normal and some older birds may be on the move too. More birds than normal will perish during a wanderjahr. Some of the more epic wanderjahr journeys have seen Barn Owls from the Netherlands, France, Switzerland and Germany find their way to Spain, covering as much as 1,650km (about 1,025 miles), and in these years a small number of Dark-breasted Barn Owls may find their way to Britain. From what is known from recoveries of ringed (banded) birds, however, distances of 300km (about 190 miles) or less are more typical.

Some surprising recoveries

Ringed (banded) birds sometimes turn up in some interesting or surprising places. Here are a few Barn Owl examples:

- A Barn Owl from southern England was found in Afghanistan. It had been ringed near a Royal Air Force base and is not thought to have made the journey under its own steam!
- Twenty months after it was caught and ringed (banded) in New Jersey, a North American Barn Owl was found in Bermuda. It may well have hitched a ride on a ship.
- A Barn Owl ringed (banded) on Humberside, on England's east coast, ended its days on a North Sea oil rig.
- After being ringed (banded) as a nestling in Essex in the east of England in July 2008, Barn Owl GC78409 was found dead in Spain that November. This was over 1,100km (about 680 miles) south of where it started life. It is suspected, however, that the bird was hit by a lorry and went to Spain on the front of it.

4 | Catching voles... mostly!

Barn Owls are superbly adapted predators, and because of their very large range and their pellets, which make studying their diet relatively easy, a lot is known about what they eat.

What do they eat?

Barn Owls are opportunistic feeders. They eat small mammals mostly – whatever they can get their talons on but especially rodents and particularly voles. In North America, Barn Owls take different prey in different regions. Voles are their main food in the north and, essentially, they eat whatever voles they can find – including Meadow Voles (*Microtus pennsylvanicus*), Prairie Voles (*M. ochrogaster*), Montane Voles (*M. montanus*) and Oregon Voles (*M. oregoni*). Elsewhere, pocket mice (*Perognathus* spp.) are an important part of the diet in the hot, dry parts of the south-west and west, and to the south and south-east Hispid Cotton Rats (*Sigmodon hispidus*) are very important food items. A whole host of other small mammals are also on the North American menu – House Mice (*Mus musculus*), White-footed Mice (*Peromyscus leucopus*), rice rats (*Oryzomys* spp.) and shrews for example, as well as the youngsters of some bigger creatures such as Brown and Black Rats (*Rattus norvegicus* and *Rattus rattus*), rabbits and hares (*Sylvilagus* spp. and *Lepus* spp.) and muskrats (*Ondatra zibethicus* and *Neofiber alleni*).

There is regional variety in the diet of European Barn Owls too. Voles, mice, small rats and shrews are the main prey of British Barn Owls, with Field Voles or Short-tailed Voles (*Microtus agrestis*) typically accounting for 50–65 per cent by weight of an owl's intake. Wood Mice (*Apodemus sylvaticus*) are almost as important and sometimes, in some areas, more important.

Voles are also the main food in much of Central Europe, though here Common Voles (*Microtus arvalis*) are the main prey. Shrews are normally the next item on a European Barn Owl's menu. For British birds, this mostly means Common Shrews (*Sorex araneus*), while on mainland Europe white-toothed shrews (which are absent from Britain) are added into the mix, particularly Greater White-toothed Shrews (*Crocidura*

Previous page: Going in for the kill. Barn Owls are experts at finding prey and will take a range of small mammals, depending on where they live. In northern Europe, voles are one of the most common prey items, but other small mammals will be eaten too.

Opposite: An everyday meal. Barn Owls normally eat on the ground or a post, or carry their victim back to the nest.

russula). Belgian, Dutch and Danish Barn Owls often eat more shrews than anything else, though Field Voles and Common Voles do occur there.

An exclusive taste for Field Voles and Common Shrews would leave Irish Barn Owls hungry, as neither of these staples occurs there. In some parts of Ireland, Barn Owls get most of their nourishment from Wood Mice, with House Mice as their secondary prey. Bank Voles (*Myodes glareolus*) have been accidentally introduced to parts of Ireland, and in these areas they are the main prey item, with Wood Mice in second place. Elsewhere in Ireland, Greater White-toothed Shrews have also been accidentally introduced and, here, the shrew ousts the Bank Vole from its prime position in a Barn Owl's diet, pushing the vole into second place. Mice are the main sustenance of Barn Owls that live in Mediterranean regions.

Other mammals are taken too. Nigel Blake saw a female Barn Owl in England attack and eat a Weasel (*Mustela nivalis*) but wasn't sure whether the Weasel was the victim of its own doomed attempt to steal the owl's prey. In their tussle, the Weasel bit the owl's foot, which swelled up, and the owl was not seen again after this encounter. Moles (*Talpa europaea*), Muskrats (*Ondatra zibethicus*) and Stoats (*Mustela erminea*), among other mammals, are also on the European prey list. Some eat bats but they are not normally an important part of the diet. Barn Owls will eat carrion but only rarely. Ian Llewellyn even saw one take bacon off a bird table!

Sometimes birds are taken, usually by disturbing them at communal roosts. Common Starlings (*Sturnus vulgaris*), House Sparrows (*Passer domesticus*), Common Blackbirds (*Turdus merula*), Red-winged Blackbirds (*Agelaius phoeniceus*), marsh wrens (*Cistothorus* spp.) and meadowlarks (*Sturnella* spp.) have all been eaten by Barn Owls, and African Barn Owls (*T. a. affinis*) will take weaver birds (Ploceidae) from their roosts. There are records of Barn Owls taking a whole host of bird species, including kestrels (*Falco* spp.), cuckoos (Cuculidae), swifts (Apodidae) and kingfishers (Alcedinidae). Peter Wilkinson, a Barn Owl worker in eastern England, has found feathers from a Common Whitethroat (*Sylvia*

communis) and a partly eaten Eurasian Skylark (*Alauda arvensis*) in a Barn
Owl nest box, and the skull of a bunting (*Emberiza* sp.) among the bones
of other prey within about 6m (20ft) of a nest. Peter believes that a small
number of Barn Owls are habitual bird killers and that it is a skill that
they seem to be able to learn – some take up the habit when voles are
hard to come by. Lusby's research in Ireland has shown that birds are a
large proportion of the autumn and winter diet of some Barn Owls. Here,
it is not unusual to find Water Rail (*Rallus aquaticus*) remains in pellets,
and in times gone by, Corn Crakes (*Crex crex*) were taken. A particularly
notable bird-killing example was reported in 1932. An island in the
Californian Pacific was host to a pair of Barn Owls that were raising their
young in a cabin. A 7.5cm-deep (about 3in) layer of leftovers confirmed
that, mostly, the owlets were being fed Leach's Storm Petrels
(*Oceanodroma leucorhoa*), which bred nearby.

Reptiles, amphibians and invertebrates are not necessarily safe either.
In Britain, Common Frogs (*Rana temporaria*) are taken, particularly in
the spring, when the frogs are on the move and vulnerable. There are

records of fish being eaten, but only rarely, and a long prey list also includes Grass Snake (*Natrix natrix*), lizards, toads, grasshoppers and crickets, beetles, dragonflies and even some molluscs. It is very rare, but not unknown, for Barn Owls in temperate latitudes to eat earthworms. Nigel Blake has seen it happen when other prey is hard to come by.

In Europe and North America, typically, the further south you go, the more varied a Barn Owl's diet. Those that live in particularly arid areas are less mammal-dependent and more insectivorous than those in wetter areas, and the diets of North American Barn Owls appear to be more reliant on just one species than those of European birds.

Different seasons, different diets

Apart from some tropical Barn Owls, this bird's diet changes through the year. Taylor found that the owls he studied in Scotland had a diet where the amount of voles eaten roughly followed the ups and downs of the vole's population. Voles were least evident in the diet between March and May, and more and more evident until the beginning of winter. For many European Barn Owls, fewer voles means more shrews, and these small invertebrate-eaters are frequently eaten in greater proportion in winter and spring. Colin Shawyer reported similar trends, though with shrews peaking in the spring and mice making more of a contribution through the winter months and into early spring. North American Barn Owls are known to vary their diet with the seasons too, one of many examples being the scarcity of pocket mice and pocket gophers (*Geomys* spp. and *Thomomys* spp.) in the winter diet. The gophers spend the winter in burrows and many pocket mice hibernate, so it is easy to see why.

Different years, different diets – the ups and downs of voles

Studies in the Netherlands, Poland and Scotland have also found, as you would expect, that the amount of voles in a Barn Owl's diet goes up and down with the voles' natural population cycles (in some parts of Europe, Field Vole numbers vary cyclically over 3–5 years). This is probably true, wherever voles are a major part of their diet, for many European Barn Owls.

Meadow Vole populations in North America vary cyclically over 2–5 years. Marti's studies in Idaho and Utah found some variation in the amount of voles being taken, but this wasn't as marked as in the European studies and lacked any predictable pattern. There is a simple explanation – the voles that these Barn Owls were eating were not from a population that was cyclic. Studies elsewhere, in different habitats, could yield different results.

Different eras, different diets

There is evidence from Britain that, over time, the Barn Owl's diet has changed. In 1913 a chimney in southern England was capped. The chimney contained many Barn Owl pellets – enough to fill three sacks and provide plenty of data on the diet of a Barn Owl in the early 20th century. These owls ate a wider range of small mammals than today's Barn Owls – 14 species were identified, of which 5 were found in similar amounts, each making up about 10 per cent or more of the diet. The British countryside of today is more uniform than it was a century ago and this could explain the difference.

Other studies in England have shown that Brown Rats (*Rattus norvegicus*) are no longer eaten in the quantities that they once were. In the 1920s and 1930s many farms were home to thriving rat populations and roughly 38 per cent by weight of a Barn Owl's consumption was Brown Rats. Forty years later their contribution was only around 17 per cent. Over the same period, Field Voles were eaten in greater quantities, plugging the gap.

Rat-catchers!

In Malaysia, Barn Owls provide a valuable pest-control service on rice farms and oil palm plantations. Plenty of owl boxes are provided and, in return, the owls help to keep the rat population under control. Barn Owls were introduced as rat-catchers to the Seychelles and Hawaii in the 1950s.

Introducing a species is a risky business and things did not go according to plan. In Hawaii, it was House Mice that ended up providing most of the Barn Owl's sustenance, and in the Seychelles the owls were happily feeding on White Terns (*Gygis alba*), a protected species, and native small mammals, rather than rats. They caused so much trouble that attempts were made to get rid of the owls, but they are still found there.

How much do they eat?

Based on pellet analysis and assuming that one pellet was produced a day, Taylor came up with a figure of about 75g (2.6oz) of food per November day for a male White-breasted Barn Owl in Scotland. That means roughly three voles a day. A similar figure has been calculated for a Dark-breasted Barn Owl in Germany. Bunn, Warburton and Wilson's work on White-breasted Barn Owls suggested a much higher daily intake of seven or eight voles (or similar-size mammals). Food requirements are higher when birds are breeding or preparing to breed. Colin Shawyer estimated that a pair of White-breasted Barn Owls might need four times as much prey during this period, their daily requirement rising from 10 pieces of food to 40. This sounds demanding, but Peter Wilkinson saw a male catch four voles in about five minutes, all within about 50m (about 165ft) of the nest site.

The North American Barn Owl is substantially bigger than its European relatives and would be expected to eat more. Calculations from pellets came up with a year-round average of about 150g (5.3oz) per day for a Californian Barn Owl and just 110g (3.9oz) per summer's day for a bird in Colorado. The first figure is very similar to the seven- or eight-vole figure (at 20g or 0.7oz per vole) for European Barn Owls, despite the greater bulk of North American birds.

The design of a killer

Barn Owls can fly without making any noise and, thanks to a remarkable sense of hearing, can locate their prey in complete darkness. Silent flight makes it easier for the hunter to hear its prey and harder for the prey to hear the hunter. Achieving it is no mean feat, and it is made possible by

three plumage features. Each wing has ten primary feathers (plus a much-reduced, tucked-away eleventh). The primaries are the biggest flight feathers and, on Barn Owls, the outermost, which is nearest the front when the bird is flying, has a comb-like front edge. This helps to smooth the airflow over the wing, cutting down turbulence and reducing noise. When a bird is flying, air flows over and under the wing surface, coming together again at the rear of the wing. Here, a soft fringe on the trailing edge of the flight feathers reduces turbulence. One more adaptation completes the package – to limit the noise of feathers moving against

The long, broad wing shown off to dramatic effect (left). The outermost primary feather has a comb-like front edge (right) which reduces turbulence and hence the noise of flying, helping the Barn Owl to live up to its reputation as a silent killer.

feathers on beating wings, the upper side of the flight feathers and wing coverts are soft and downy. Barn Owls can fly in silence but they don't always. Take off and landing may produce some wing noise, and if the feathers are sodden or in poor condition, the silent hunter might not be so silent.

Barn Owls have lightweight bodies and long, wide, rounded wings, a combination that results in low wing loading and plenty of lift. They can fly at slow speeds, twisting and turning without falling out of the air, and can carry relatively heavy victims back to a hungry family with little effort.

Unlike many owls, a Barn Owl's eyes are small and dark. Their night vision is good, better than ours but probably not as good as a Tawny Owl's (*Strix aluco*). At night, they can see buildings and trees well enough to avoid flying into them, and back in 1952 Curtis proved that Barn Owls can do this when it is so dark that a person could see absolutely nothing. To make the most of the light that is available, the lens and cornea are large, so more light gets into the eye, and the retina is laden with 'rods' (cells that work well in low light). Because there are so many rods, there are not many 'cones' (the cells that see colour), and a Barn Owl's ability to see colour and detail is probably not particularly good – they have been known to land very near, or even on, people!

Their eyes are tubular and, unlike ours, cannot be moved in their sockets. They provide binocular vision but their field of view is only about 110 degrees. To make up for these limitations, a Barn Owl can turn its head through an impressive 270 degrees or so, to the left or right. Like all birds, Barn Owls have a third eyelid, the nictitating membrane. The Barn Owl's is opaque and tougher than most other birds and can be pulled across to provide protection when hungry owlets are being fed, feathers are being preened, and as it grabs or eats its prey. When its eyes are covered in this way, a Barn Owl can resort to touch, using the bristles around its beak to feel its way, ironically perhaps, like a vole might use its whiskers.

Opposite: Compared to many other owls, the Barn Owl's eyes are small and dark. Their night vision is better than a human's but probably not as good as a Tawny Owl's.

A Barn Owl's sight is good but its hearing is better – it can pinpoint the position of its prey using hearing alone. Payne and Drury proved this in 1958. They worked with captive Barn Owls in a light-tight room with dead leaves on the floor and the owls caught live deer mice in absolute darkness. The heart-shaped face is much more than ornamental – it functions as two parabolic reflectors, akin to those used by sound recordists, which amplify sounds and direct them into the ears. The ears are located underneath the facial disc, near the eyes, and the ear openings are covered with a flap of skin. The flap on the left is about 2–3mm (about 0.1in) higher than the one on the right. On the right-hand side of the face, sounds are channelled *down* to the ear, the fleshy part of which points slightly upwards. On the left-hand side of the face, sounds are channelled *up* to a slightly downward-pointing ear. There is asymmetry

This White-breasted Barn Owl is cleaning its claws. Note the tactile bristles around the beak – these are used to help the owl feel its way when its eyes are covered by the nictitating membrane.

but it is not as marked as in many other owl species. Note, too, that the Barn Owl is able to change the shape of its facial disc, from a classic heart shape on a relaxed bird to a much rounder shape when the bird is more vigilant.

As an airborne hunter, a Barn Owl needs to know the horizontal angle to its prey – how far left or right it is – and the angle of elevation between itself and its victim. The differences between its two ears enable it to interpret sounds in such a way that it can do this with great precision. Prey in front of the bird has little chance, its position being determined with supreme accuracy. It is much less accurate if the prey is further to the left or right, or positioned so that the angle of elevation from the owl's ears to the prey is more than 45 degrees.

A Barn Owl uses its hearing to do more than locate the source of a noise. It is particularly good at detecting noises made by small mammals when they move among the leaves or grass, when they are eating and, of course, when they call. It can pick out a particular noise even when it is among others of a similar frequency, but there are limits, and excessive traffic noise may force a bird to hunt further away from the road. Colin Shawyer thinks that a Barn Owl's hearing is such that it can tell the difference between a moving vole and a breeze rustling the grass. It may even be able to tell a shrew from a vole. Their hunting is opportunistic, however – they are not selecting one over the other. Its phenomenal hearing even allows a Barn Owl to take prey from under snow. If more information is needed it

The Barn Owl's facial discs amplify sounds and channel them towards the ears, thus playing a vital role in hearing. The owls can modify the shape of the face: a rounder face indicates a vigilant state, a strongly heart-shaped face is more relaxed.

can rock from left to right or turn its head from side to side to provide
data from different positions. Whatever data it collects is analysed in the
auditory part of its brain. This has the processing power of around 95,000
neurons, whereas a day-flying crow has just 27,000.

Grasping prey in long vegetation or through snow is made easier by
the Barn Owl's surprisingly long legs,
long toes and very sharp talons. Three
toes point forwards and one to the rear,
though the outermost forward-facing toe
swings to the rear so that prey is grabbed
with two toes on the left of the victim
and two on the right. Lumpy pads
(papillae) on the toes make the grip even
more secure.

Hunting techniques

Barn Owls are nocturnal hunters but, in
Europe at least, it is not unusual to see
them out and about an hour or two
before the sun sets and after it rises, with
different individuals adopting different
habits. Daytime hunting does occur, and
Peter Wilkinson, who works with Barn
Owls in the east of England, believes

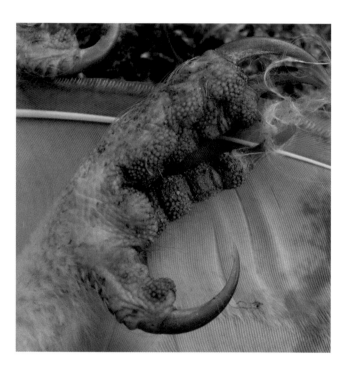

The structure of the feet and legs enables the owls to take prey from long grass and even from under snow. Sharp talons and a rough surface on the toes provide a secure grip and make it unlikely that prey will be able to wriggle free from the hunter's grasp.

that food stress is the main reason for it. This could be because wind or
rain has made hunting difficult, or it could be young birds that have
dispersed and are struggling to find food in the winter in suboptimal
habitats. Alternatively, it could be a male working hard to find food to get
a female into breeding condition or to feed young when summer nights
are just too short. Northern European birds are different – in the north of
Britain, for example, White-breasted Barn Owls routinely hunt in
daylight, and in darkness, all year round. *Birds of North America* paints a
more nocturnal picture, with hunting normally starting after dusk and

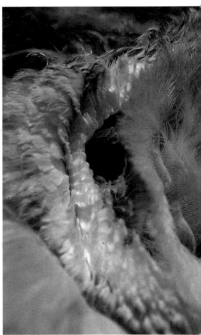

Left: The Barn Owl's long legs, not obvious in life, can be seen clearly in the skeleton. Note, too, the long bill and long toes and talons.

A vital part of the Barn Owl armoury: one of its ears. Experiments have shown that Barn Owls can locate prey in complete darkness by using their highly developed sense of hearing.

finishing before dawn, though some North American Barn Owls do sometimes venture out in daylight.

They hunt using their eyes and their ears and sometimes use artificial light to assist them, making the most of roadside lighting and illuminated farm buildings, for example. Nigel Blake has seen Barn Owls position themselves with street lighting in front of them, so that they can use the light to hunt small mammals crossing a wet road. Several hunting techniques are employed by this deadly owl.

Quarter and drop

A quartering Barn Owl seems to float through the air, jinking, ascending, turning, descending, shifting to the left or right, tumbling and perhaps hovering, as it searches for a meal. When food is located, the owl may drop on to it from directly above, plunge forwards towards its prey or

engage in various aerial acrobatics to get talons on meat. The owl dives head first, pulling its head up, closing its eyes and bringing those long legs forwards, with toes spread, only just before it seizes its victim. Once on the ground, the bird opens its wings for steadiness. Some use a slightly different aerial technique, moving between different spots that are checked out from a hover.

Quartering flights are low level, often between 1.5 and 4.5m (4.9 and 14.8ft) off the ground. Shawyer's work suggests approximately 2m (6.6ft) as a typical quartering height, while Taylor estimated the theoretical optimum height as about 3m (9.8ft), which he believed matched the behaviour of birds in the wild.

Perch and plunge

Sitting on a suitable perch, waiting for up to eight minutes and then taking the plunge towards unfortunate and unsuspecting prey is another common hunting method. Sometimes a perched bird will lean forwards so that it is looking at the ground directly below. Perch-and-plunge hunting is seen more in the winter than in the summer, perhaps because it is more energy

Hunting flights are typically low to the ground, and may include a variety of flying styles as the owls search for food.

efficient than other methods. Taylor suggested another possible explanation – it could be simply that winter hunting habitats have plenty of perches and summer ones do not. A variation on this technique involves loitering on a track or road and waiting for a meal to cross.

Bush whacking

This is not a common hunting method but is one way of getting roosting birds out of cover. The owl flies close to a tree or bush and strikes it with its wings. Typically, this is a tactic employed at the end of the day, when roosts gather, especially in the winter.

Killing and eating

Not every hunting attempt is successful, especially for younger birds, and an owl that missed might be seen chasing its prey on the ground. When they are successful, to protect their catch from would-be thieves, a Barn Owl will mantle its prey. This simply means spreading its wings so that the prize cannot be seen and stolen by piratical competitors, the Common Kestrel being a good example.

Prey is either squeezed to death in the owl's foot or sent on its way with a lethal peck to the rear of the head. Colin Shawyer has found that Wood Mice, which are particularly lively creatures, are usually despatched with more than one peck to the back of the head. Small mammal prey goes down in one piece, head first, including, on occasion, prey as large as a small Brown Rat. Anything too big to go down in one is torn up and swallowed in smaller pieces. Birds and larger mammalian prey are often 'prepared' before being eaten. This involves removing the head and, for birds, pulling out the bigger feathers or removing the outer part of the wings. Barn Owls normally take their meals on the ground or a convenient post, though they may, of course, ferry a catch back to the nest, using either a single foot or both of its feet to haul it.

Food caching

Using food to win a female is hardly an original approach but it works for the Barn Owl. Normally, a bachelor male will create a grisly hoard of dead animals at a would-be nest site, much of which may be rotting. Iain Taylor found a hoard in Scotland that totalled 136 victims. It may

Left: A successful outcome for the owl, if not for its prey.

Above: No sharing allowed. Barn Owls will jealously guard their prey from competitors by 'mantling', spreading the wings to hide the victim from other birds that would appreciate an easy meal.

be that the hoard is used to impress any female that succumbs to the male's screeching song and ventures into the nest site.

Males that have found a female often cache food and will draw from this larder as part of the mating ritual. Barn Owl foreplay normally includes the male giving the female a gift from the cache. She does not eat this offering there and then but usually grips it in her bill during coition and gulps it down afterwards. If she is not hungry, she puts the item to one side instead.

A cache can also help to supply the female's nutritional needs when she is making eggs, when extra energy and protein are essential. She does not hunt for herself at this time and, apart from the resources that are already stored within her body, must rely on food brought to her by the male, either straight from the hunt or from a cache (typically at his roost).

Colin Shawyer found that Barn Owls cache food from a little before egg-laying until there are young chicks in the nest and suspects that the male may choose smaller items from the cache when the chicks are small. The male is particularly intent on bringing in food when the eggs are being laid and can supply more than the hen needs, which results in food accumulating at the nest. This also happens during the first few weeks after hatching, but once they pass the three-week mark, the young birds' appetites go up. Thirty or more food items can accumulate at a nest, with Wallace's study of Barn Owls in Michigan reporting an amazing 189 items. Why Barn Owls cache food at some times and not others is not well understood.

Male Barn Owls may bring more food to the nest than the developing family can eat, leading to an accumulation of prey.

Waste disposal – pellets

Pellets are the undigested parts of a bird's food that are lumped together and then coughed up. A Barn Owl's stomach is not as acidic as that of many birds of prey and, as a consequence, it is not good at digesting bones or fur. This means that a Barn Owl pellet can contain skulls and other prey remains that are in good condition and not difficult to identify. Their pellets, therefore, tell us a lot about this bird's diet. They also help scientists work out what small mammals are in an area and can lead to exciting discoveries. In 2008 John Lusby was researching Barn Owl diet in Ireland as part of a PhD thesis. He was particularly interested in the importance of the introduced Bank Vole as Barn Owl food and pellet analysis was a key part of the work. Ten pellets from a single site in Tipperary contained something completely unexpected – the skulls of 53 Greater White-toothed Shrews, a species that had never been recorded in Ireland. It is possible that the shrew was introduced with plant material imported from the Netherlands to a nearby horticultural centre but this has not been proven. Since then, the shrews have been found in pellets from a wider area and a separate population has been discovered in County Cork.

Barn Owls are not good at digesting bones or fur, so their pellets contain remnants, such as the skull seen here, that can help researchers identify prey.

A daytime roost site in a building is probably the best place to find Barn Owl pellets. Elongated, vertical lines of chalky, white guano beneath the roosting spot may well point you in the right direction, with, potentially, tens of pellets awaiting discovery on the ground underneath. If the site isn't used very often, fewer pellets will have been regurgitated there. Barn Owl pellets vary in shape and size – from very small, round examples to much larger, elongated specimens with rounded ends. Some examples are shown opposite. A bigger than normal meal can lead to a bigger than normal pellet, with a rounded front end and a tail at the rear. Pellets can reach a gagging 140mm (5.5in) in length, though this is not typical. An average pellet, according to Colin Shawyer, is 42mm (1.7in) long and 26mm (1.0in) wide. Work by David Glue published in 1970 gave a normal range of 30–70mm (1.2–2.8in) long and 18–26mm (0.7–1.0in) wide. *Birds of North America* does not give any dimensions for pellets produced by North American birds.

Two pellets a day is regarded as the norm for European Barn Owls. Work in North America gives an average of less than two. Captive Barn

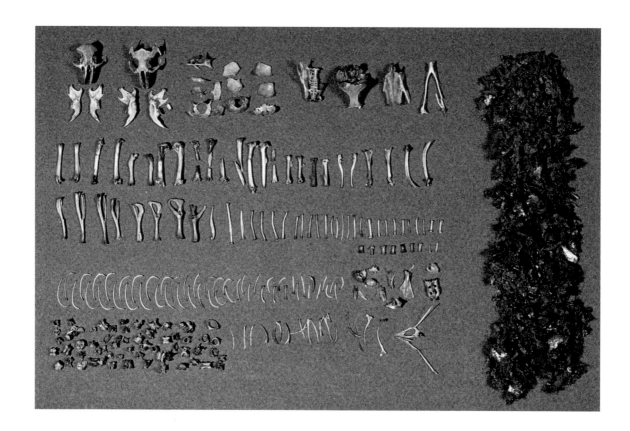

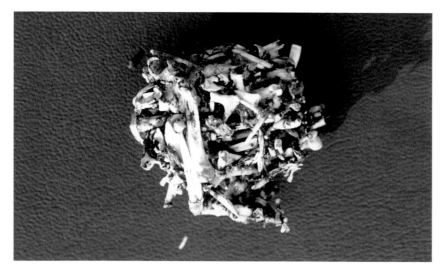

Above: The contents of a single pellet. An average pellet is over 40mm long.

Left: Although a diet of mammals is usual, Barn Owls will also eat other prey items. In this case, the owl had been feasting on frogs, resulting in a very different pellet from those most commonly seen.

Owls often produce a pellet when they think they are about to eat – pellet production has been triggered by the sight of a live vole and by showing them food. In the wild, Barn Owls frequently eject one pellet a little before leaving the roost to hunt and another at a night-time roost. In 1972, Smith and Richmond worked out that a Barn Owl needs at least six and a half hours to make a pellet.

Poking around in a pellet is a fascinating experience. A pellet from an adult Barn Owl usually contains bits from between three and six mammals. Fewer and more are possible, of course – one that contained shrews held the remains of 14 animals. Occasionally, Barn Owl pellets are made up mainly of plant material – there are records of pellets that consist mostly of grass or leaves, and even stones have been found in them. Exactly why Barn Owls are eating these things is unclear.

How to age a pellet

Once a pellet has dried out, it is tough, compacted and quite hard to pull apart. Its appearance and consistency change over time – the table below is a rough guide to help you work out the age of a Barn Owl pellet.

If a Barn Owl has been eating frogs its pellets look quite different – according to Colin Shawyer, they look like white golf balls.

What do barn owl pellets look like?

AGE	DESCRIPTION
Fresh out of the owl	Soft, squidgy and steaming!
4–5 days	Still damp and soft
1–2 months	Dries out and goes black and shiny
2 months plus	Black slowly wears off. Turns progressively grey/brown
6–12 months	Wholly grey/brown
12 months plus	Clothes moth caterpillars hatch in the pellet, eat the fur and leave the bones

5 | Home and family

An area defended by a bird is usually called its territory. An area that a bird lives in but doesn't defend is its home range. Barn Owls are not very territorial, or at least they do not defend a particularly large area. In the breeding season, the home range of a British Barn Owl averages roughly 3km2 or 300ha (about 740 acres). Nigel Blake followed a Barn Owl in Bedfordshire, England, that was hunting a circuit that began and ended at its nest site. The circuit totalled about 11.6km (7.2 miles), suggesting a home range that was much larger than 3km2. John Lusby's radio-tracking work has shown that some Irish Barn Owls have home ranges that are substantially larger than the British average, with birds venturing as far as 6km (3.7 miles) from the nest during the breeding season. This reflects a lower availability of prey, and where Bank Voles and Greater White-toothed Shrews have been introduced, Irish Barn Owls have home ranges that are closer in size to those of typical British birds. Nick Askew's work in the UK found that birds with a smaller home range produce more young, presumably because of the efficiency with which food can be found and delivered to the nest. The home range is where they hunt, roost and nest and is an area that the occupants know well, often flying the same route to a preferred hunting patch. They certainly defend the area closest to the nest, up to somewhere between 5 and 100m (16 and 328ft) away from it, but beyond this there seems to be little or no territorial behaviour.

Work in Germany has produced an average size for a breeding-season home range of 188ha (465 acres), and studies in the United States have come up with a very large range of figures, some smaller than in Britain and others much bigger.

The German research found that home ranges were larger outside of the breeding season and that birds also hunted further from the nest site then. These home ranges covered between 363 and 465ha (897–1,149 acres) and the birds were flying an average of 2.2km (1.4 miles) from the nest site to find food. This is over three times more than the distance they were flying during the breeding season. Taylor's work in Scotland found that birds were hunting as much as 4.5km (2.8 miles) from the nest in the

Previous page: Most Barn Owls pair for life and are monogamous, although some are known to develop more than one partnership. If one of the pair dies the other will usually seek another partner.

winter. In the summer, 2km (1.24 miles) was the limit, though almost 90 per cent of the sightings were a kilometre (0.6 mile) or less from the nest. It could be that Barn Owls need to travel further to find enough food or just that they can take advantage of different hunting areas because food does not have to be airlifted back to hungry dependents. Colin Shawyer also suggests that they may travel further simply to find a roost site that will enable them to survive the colder winter months.

Home ranges are not necessarily mutually exclusive and it is not unusual for Barn Owls from more than one pair to hunt in the same area. Colin Shawyer found that they normally worked the area at different times, though even when they were no more than 50m (about 164ft) apart there were no major conflicts. In March 2010 Nigel Blake watched a total of seven Barn Owls hunt over the same field in Norfolk, England. Where food is harder to come by, conflicts are more likely, however.

Dutch Dark-breasted Barn Owls at their nest site.

Nest sites

Many readers will think of the Barn Owl as a bird that nests in buildings, nest boxes and holes in trees. These sites certainly are used, but so are caves and holes in the ground, riverbanks and rock faces (including sea cliffs). Barn Owls are also not averse to setting up home in the nests of other birds. Taylor found Barn Owls pushing Western Jackdaws (*Coloeus monedula*) out of their nests in Scottish chimneys, and they have also been known to move into Western Osprey (*Pandion haliaetus*) nests, sensibly perhaps, when the Ospreys have moved out. In Africa, some Barn Owls take advantage of the huge, stick-built, domed nests of the Hamerkop (*Scopus umbretta*) and others have made themselves at home in the Sociable Weaver's (*Philetairus socius*) nest, another large and equally remarkable structure.

Barn Owls nest in a wide range of man-made structures, including barns and other farm buildings, silos, roof spaces in derelict houses, castles, church and cathedral towers, windmills, bridges, wells, water tanks and hay or straw stacks. John Lusby knows of nests in Ireland in the

Above and opposite: Barn Owls will nest in a wide variety of sites, including hay and straw stacks and trees as well as many man-made sites and even riverbanks and cliffs.

chimneys of occupied houses, which were blocked by Western Jackdaw nests. On mainland Europe many opt for man-made nest sites, with church towers a particular favourite. Here, nests in trees are rare, but they are used much more frequently in Britain, and particularly in England, with oaks (*Quercus* spp.) and Ash (*Fraxinus excelsior*) being the preferred species. Elms were once an important nesting tree for Barn Owls, but sadly many have now been lost after Dutch elm disease struck in the 1960s. An analysis of British Barn Owl nest sites from 1982 to1985 found that 29 per cent were in trees and 69 per cent in buildings. There is regional variation though – Jeff Martin's work in Suffolk (a county in eastern England) found that over 80 per cent were in trees, while Iain Taylor's Scottish study found that over 85 per cent were in buildings! Taylor's work also found that buildings were used more often simply because there were more buildings than trees – the owls were not actively choosing them instead of tree sites.

North American Barn Owls seem to opt for natural sites more than their European counterparts. Colvin's New Jersey study (published in

1984) found that about 50 per cent were tree nesters, with Silver Maples (*Acer saccharinum*) accounting for 66.7 per cent of the tree sites and the American Sycamore (*Platanus occidentalis*) for another 28.4 per cent.

Where nesting opportunities in buildings, trees and haystacks have been lost, nest boxes are increasingly important for the Barn Owl's future. Nest boxes can be sited in or on the outside of buildings, on trees or on poles. In Colvin's study 31 per cent of his New Jersey Barn Owls were breeding in nest boxes, but that is nothing compared to parts of Utah, where Marti (1994) found that nearly all of the Barn Owls were using nest boxes. Nest-box use in Britain has increased and, in 2010, Colin Shawyer believed that about 70–75 per cent of the nation's Barn Owls were nesting in nest boxes.

The perfect nest site?

The best Barn Owl nest sites are dark spaces that offer protection from the wet, with an entrance hole that is a minimum of 15cm (6in) wide and with more than 30 x 30cm (1 x 1ft) of floor space. A wide range of cavity sizes are used and the cavity is often much bigger than this prescribed minimum. Barn Owls, especially the North American subspecies, do not like sites where there is too much human activity. This seems to be less of an issue with at least some British Barn Owls, where nesting may take place despite nearby activity. Peter Wilkinson tells of a nest in an outbuilding in Hertfordshire, England, that went undetected until the chicks were five weeks old, despite the daily activities of people within 5–10m (about 16–33ft) of the nest.

Tree nesters usually opt for trees that stand alone in hedges and fields and use a large hole or cleft in the trunk, a cavity in a rotting stump or broken-off trunk or perhaps nest in a hole in the trunk of a pollarded tree. Most of the holes used by British Barn Owls are 3.5–6m (11.5–19.5ft) off the ground, though they can be more than 10m (33ft) up and sometimes are only just off the ground. The height of the entrance hole can be deceptive – in a hollow tree the owls may descend up to 6m (19.5ft) to reach the floor of the nest. North American birds tend to go for higher

A pole box. Boxes can also be mounted on trees and in or on buildings.

holes, perhaps because of marauding Northern Raccoons (*Procyon lotor*) – in New Jersey, Barn Owl holes averaged 6.8m (22.3ft) up, with the lowest at 2.5m (8.2ft) and the highest at 14.3m (46.9ft). Peter Wilkinson has found that Barn Owl nest boxes do not need to be positioned more than 3–3.7m (10–12ft) off the ground. He has seen birds occupy an indoor box about 2m (6.6ft) or so off the ground and an outdoor box roughly 2.5m (8.2ft) up.

Who builds the nest?

Mostly, neither bird does any real building, with the exception of those in New Mexico and Colorado that use their feet to excavate a burrow in a river or arroyo bank (an arroyo is a creek bed that only carries water after substantial rainfall).

In many nests pellets are used to make a bed for the eggs to rest on. The female will tear the pellets up as part of the process. Later, she may pull pellets in towards her, encircling herself in the remains of previous meals. Sometimes, however, Barn Owls will simply deposit their eggs on a wooden floor.

Co-habiting

It is not unknown for Barn Owls to nest in close proximity to each other. In 1984, three females nested successfully in the roof of one house in southern England. North American Barn Owls have nested communally too and when the African subspecies moves into a Hamerkop nest it may do so en masse. Back in Britain, and back in the 19th century, Barn Owls sometimes lived in 'owleries', with several tens or more sharing a common roof space.

Communal living extends to other species too. Barn Owls will certainly live alongside Western Jackdaws. One particularly notable example comes from southern England, where, in 1988, a hollow in an Elm tree was home to a pair of Barn Owls. There was only one way into the cavity and this was occupied by three pairs of Jackdaws. Sometimes, though, Jackdaws are direct competitors for nest sites and may limit the

number of Barn Owls that are able to breed. Some Barn Owls choose dovecots as nesting sites, with doves as very near neighbours.

Barn Owl boxes may be built with one species in mind but they are desirable residences and are sought after by other species too. Some boxes have two compartments and Common Kestrels and Barn Owls sometimes share a box, with the falcons in the upper compartment and the owls down below. There could be Stock Doves (*Columba oenas*) waiting hopefully just outside too. Kestrels have also been known to nest at the entrance, so the Barn Owls have to walk past them to get in. Colin Shawyer tells of a two-compartment box that was occupied by a pair of Little Owls (*Athene noctua*) as well as the intended residents. Even more remarkably, in 2009 in Lincolnshire, England, Barn Owls and Kestrels shared the very same space in a nest box and both were successful.

How close are the neighbours?

As we have just seen, sometimes Barn Owls are very close to other Barn Owls. In Europe and North America this is not the norm, though American birds sometimes breed in loose colonies.

A survey of Britain and Ireland in the 1980s found an average of just over two pairs of Barn Owls in each 10km square (an area of 100km2) where Barn Owls occurred and a peak of eight pairs in 100km2. This is the equivalent of one pair per 12.5km2. In excellent habitat in Germany, however, densities as high as one pair per 4.3km2 have been recorded, and Glutz von Blotzheim's 1962 publication quotes a figure of one pair per 0.25km2.

Providing nest boxes makes a difference, perhaps most notably in Malaysian oil palm plantations. Here there is plenty of food and lots of nest boxes and a single 10km square can support over 250 pairs, which equates to one pair every 0.4km2.

Cambridgeshire farmland in the east of England may be less in need of rodent control than parts of Malaysia but boxes make a difference here too. In 2007, a farm of roughly 2.5km2, which is well endowed with Barn Owl boxes, supported seven pairs of Barn Owls, a density very

Previous page: A White-breasted Barn Owl peering out from a tree. Entrance holes in trees can be some distance from the nest, which may be several metres below.

similar to that of a Malaysian oil palm plantation. As well as the owls, the boxes also housed five pairs of Kestrels.

Faithfulness

Barn Owls in Britain are very faithful to their home ranges. This has been confirmed by ringing (banding) recoveries and by more intensive, localized work. Taylor's Scottish study found that 99.3 per cent of 137 males and 95.1 per cent of 150 females stuck with the same nest site for two successive breeding seasons. Of the eight birds that moved, six did so after losing their partner, two moved on from a bigamous relationship with a male and none of them moved more than 8km (5 miles). British Barn Owls often demonstrate remarkable site loyalty and may stay put even when their home patch habitat is seriously degraded.

Barn Owls in other parts of Europe seem to be less loyal to a particular home range. Ringing recoveries from the Netherlands and Germany reveal that 50 per cent or more of the adults move over 10km (6.2 miles) and 50km (31 miles), respectively, from one breeding season to the next. In North America some Barn Owls are very site-faithful and others are not. Marti's work in Utah found a high degree of site loyalty, while studies in New Jersey found 38 per cent of the females and 18.9 per cent of the males changed nest sites from one year to the next. Having said that, one nest site may be used time and time again but by a succession of different Barn Owls. A higher degree of site fidelity in males has also been supported by limited data from France, though there is an old record of a female using the same nest site for nine years.

Roosts and roosting

A Barn Owl home range includes hunting areas, a nest site and a number of other roosting sites. Studies in New Jersey found up to 7 roosting sites, while Shawyer's work found that Scottish Barn Owls were more conservative, with 3–5 regular roosting sites, and Taylor found just 1–3. Barn Owls roost in buildings, holes in trees, riverbanks or rock faces, in gaps in piles of hay bales and in nest boxes. On mainland Europe church towers are often used and North American Barn Owls take advantage of disused storage silos. Some find roosting shelter among the foliage of conifers or Holly (*Ilex* spp.), and in North America Barn Owls are known to have roosted on the ground among crops. Winter roosts are sometimes shared with other Barn Owls – Taylor found as many as four females or two males in residence at a single roost.

Roost sites are used at night, for rests between foraging sessions and, of course, when the birds are off duty during daylight hours. They can be very near to the nest – just a few metres away, under the same roof – or much further away. Taylor found roosts as much as 3km (1.9 miles) away and that the females stayed closer to the nest than the males. Work in New Jersey showed American birds travelling even further to their roost sites – as far as 5km (3.1 miles) was normal and up to 8km (5 miles) was not unknown.

Colin Shawyer looked at the roosting behaviour of ten pairs of Scottish Barn Owls and found that the male and female may or may not roost with each other outside of the breeding season and may use the nest site as a roost or roost somewhere else. Barn Owls that are sleeping apart begin to roost together again in the winter before the next breeding season begins. Shawyer found that, in Scottish Barn Owls, the male then roosted with the female until the eggs were about to hatch, at which point he moved out but did not go very far, roosting under the same roof or very nearby and normally within sight of the nest. About a month after the first owlet hatched, the female moved out and roosted between 500m (0.3 mile) and a kilometre (0.6 mile) from her offspring. At this point some males took to roosting with their partner again; others moved to a different roosting site (as much as 1.5km [0.9 miles] from the owlets) and some just stayed put.

In the United States, Marti found that, among his Utah study birds, it was not unusual for the male and female to be sharing a nest-site roost from November and, similar to Shawyer's Scottish birds, most males had moved out when there were chicks in the nest. Baudvin's work in France, however, found that, while most males are at the nest prior to egg-laying,

Opposite: Like nest sites, roosts can be established in a wide range of locations, including church towers and other buildings.

only 31 per cent are there when the eggs are being laid.

Taylor's work showed that most females are at their nest sites about six weeks before the first egg is laid and they are all in *situ* about a week and a half in advance. Males arrived later but most were also roosting at the nest site about a week and a half ahead of the first egg. When the owlets were around 7–10 weeks old, most of the adults had moved out of the nest, though a minority of females were still roosting there.

A mate and mating

Mostly, Barn Owls are monogamous and pair for life. 'Divorce' is not unheard of but it is rare. There are records from Switzerland of males switching partner between clutches in one breeding season. Marti's Utah study noted 'divorce' in just one pair in 200. Should a partner die,

A developing family in an unusual home.

however, the other bird will find a new partner.

There are some recorded instances of polygamy. A one male–two female arrangement has been recorded in England, Scotland, Spain and the United States. This can be a very close arrangement, with the two females in the same nest box, or more distant, with the two nest sites up to 4.5km (2.8 miles) apart. Peter Wilkinson suspects that bigamy has become commoner in his study area since about 2007. A one female–two male arrangement is known from the work of Roulin and others in Switzerland. Here the females abandoned their first brood and found a new male to breed with, up to 10km (6.2 miles) away, with no adverse effect on the deserted young.

Barn Owls will breed before their first birthday, though some males hold fire for another year. In north-west Europe Barn Owls lay their eggs mostly between April and August, though early birds can start late February and late sitters can still be on eggs in early October. In Britain, egg-laying doesn't normally start before the end of April or the beginning of May. North American Barn Owls start at different times in different places. Texan birds start from January to October; typical Californian Barn Owls lay their first egg in February; and birds in northern Utah, according to Marti, start laying between 4 March and 12 April, with the earlier dates being achieved after milder winters.

The breeding season starts between six and eight weeks earlier. Males that are not already roosting at the nest site will visit it and increase the amount of time that they spend there. They get noisier six or seven weeks before the first egg makes an appearance. The male's screeching song rips through the air, up to 250 times a session, as he shouts out his presence, typically from somewhere close to the nest, or would-be nest. He is telling other males to keep clear and asking females to come closer. Short display flights near the nest reinforce the message, as he flies with pronounced, rigid wingbeats, screeching in mid-air. An amorous male will chase a female through the air, a little higher but staying close to her tail through twists and turns, each bird screeching as they go. This can be an energetic chase or a more relaxed affair. A male may also use 'moth-flights' to impress a perched and watching female. This involves a brief hover with legs hanging, right in front of a perched female.

Screeching is also a prominent part of the male's nest-showing displays, when he flies to and from the would-be nest, again and again, screeching again and again. Sometimes a male will stay at the nest site and call, or add a display flight to the routine, calling and changing course repeatedly during a short sortie above his home range. During nest-showing a male may also use a purring call to bring in the female.

A pair will mate surprisingly often. They will be mating while they are checking out nest sites, with just a few minutes between couplings, and can still be at it when there are chicks in the nest, though it happens less and less often during incubation and thereafter. The more they have mated, the bigger their clutches tend to be. As discussed in the previous chapter, food plays an important part in courtship and mating. It is not unusual for a courting male to be caching food at the nest, and mating is often preceded by a small mammal gift from the male, which the female holds in her bill during mating and either gulps down afterwards or leaves on one side. The male's greeting call – a squeak and some twitters and chirrups – tells the female that he is there. She snores and purrs and the male comes closer, uttering another greeting call or twittering. The gift is given and the female stoops, encouraging the male to mate with her. Either bird can initiate proceedings by crouching in front of their partner, with the female using subdued, accelerating snores and/or purring calls as part of the invitation.

During the act itself, the male holds the back of the female's neck in his bill, spreads his wings for balance, makes high-speed staccato squeaks, may snore a little and thrusts his cloaca to hers repeatedly, connecting once every two or three seconds. Copulation can last from ten seconds to about a minute and while it is going on the female's snores usually get louder and louder. When it is over the male might screech once or twice and then snooze, with the female preening him. Copulation normally takes place at the nest site or nearby, often as a prequel to the evening's first foraging trip, as well as when the male comes back with food and at other times.

As the breeding season moves towards egg-laying, the female does less and less and does not venture far from the nest. She will snore or purr to encourage the male to bring her food, and when he does, they usually mate. The female gets heavier and heavier, preparing her for egg-making and the demands of incubation. She becomes nest-bound about a week and a half before the first egg is laid. The male must provide all her nutritional needs, either freshly caught or from a cache.

Barn Owls mate frequently. The male grips the female's neck with his bill and spreads his wings for balance.

When the eggs are laid the parents get quieter. A Barn Owl egg is elliptical, with an even, dull white surface. Barn Owls are not fastidious housekeepers, however, and these clean, white eggs can become very soiled amidst the debris of the nest site. European Barn Owl eggs weigh between 17.5 and 22g (0.6–0.8oz), with Dark-breasted Barn Owl eggs averaging 39.4mm (1.6in) long and 30.9mm (1.2in) wide and White-breasted Barn Owl eggs 42.1mm (1.7in) long and 31.8mm (1.3in) wide. Marti measured the eggs of North American Barn Owls in Utah. Not surprisingly, this larger subspecies has larger eggs, measuring 43.8mm (1.7in) long and 33.4mm (1.3in) wide and tipping the scales at 25.2–28g (0.9–1.0oz). For its size, however, the Barn Owl lays relatively small eggs. This could be a way of spreading its egg-making resources so that more eggs, and therefore more owlets, can be produced.

A typical clutch contains 4–7 eggs, laid 2 or 3 days apart, though clutches as small as 2 eggs are known and, in the wild, as large as 16. A captive bird laid 26 eggs in just under 9 weeks. Some Barn Owl workers suspect that very large clutches are actually two females' eggs, with just one bird incubating. In Taylor's Scottish study, the better-fed, heavier females produced bigger clutches. Food influences breeding in other ways too. European studies have found that the more field voles there are the bigger the clutch, and, in the Netherlands and Scotland at least, fewer Barn Owls breed during dips in the vole cycle. Shawyer describes a partnership between prey availability and the Barn Owl's food requirements, with Field Vole numbers rising to their maximum just when the season's youngsters are learning to find their own food.

In many areas of Europe, clutches get bigger until July and then get smaller again. Scottish data are different, with Barn Owls here laying

Previous page: A family of five. Most clutches contain between four and seven eggs. Clutches are often larger when food is plentiful, and some pairs will have two or more broods in a single season.

Eggs are relatively small considering the size of the adult. The clean white surface does not stay pristine for long and will quickly become soiled among the detritus of the nest.

their biggest clutch at the beginning of the season and then smaller clutches. A study by Johnson in Norfolk (eastern England) concluded that Barn Owls that nested in buildings or trees laid smaller clutches than those in nest boxes.

Second, third and fourth broods

Some pairs go on to lay a second brood, and Peter Wilkinson believes that this is commoner than people realize. Almost four months will pass between the first egg being laid and the young gaining their independence, so double-brooders must start early. Paddy Jackson, one of Peter's colleagues, has calculated that, as a rule of thumb in eastern/central England, Barn Owls need to have their first egg in the nest by 20 April at the latest for there to be any realistic chance of a second brood. Studies in Europe have found that, on average, the second clutch is started 90–100 days after the first. Double broods are commoner in good vole years, when food is plentiful. The importance of a good food supply has been shown in England, where there were more second broods among Barn Owls that were artificially provided with additional food than those that were left to their own devices. When pairs do go on to a second brood, this can be in the same nest as the first brood or somewhere else, and the second brood may be initiated before the young from the first have fledged. A pair in Utah split the work two years running, with the male feeding the young of the first brood while the female incubated the second clutch elsewhere. As an alternative approach, Roulin found that 46 per cent of double-brooded females abandoned their first brood and found a new mate for their second attempt. Only 4 per cent of males did this.

Marti recorded second broods among 11 per cent of his Utah study birds, while, in California, 56 per cent had a second go. More remarkably, work in south-east Spain by Martinez and Lopez recorded *third* broods in almost 13 per cent of the breeding population. Roulin, however, did not find any third broods in over a decade of studying Swiss Barn Owls but it is known to have occurred in Germany. In

Eating family members

Cannibalistic Barn Owls were brought to the attention of the British public in 2007 when viewers of the BBC's Springwatch programme witnessed a Barn Owl chick eat a live sibling. The father had gone missing and it seems that one of the owlets was determined not to go hungry. In a different example, an adult female is known to have killed one of her own very young chicks as food for herself and, in another case, as food for one or more of its siblings. The frequency of Barn Owl cannibalism has been described as both 'occasional' and 'frequent'! A 1999 study of Swiss Barn Owls by Roulin and others found that 10 out of 60 dead owlets had been cannibalized. These owlets were not killed to be eaten – they had died of some other cause and that has often been said to be the case when cannibalism occurs. Clearly, it is not always the case.

Zimbabwe, the African subspecies, generously nourished by a plague of mice, has even managed four broods in less than a year.

Incubation and hatching

As egg-laying approaches, the female acquires a brood patch. This feather-free area on her belly has a very good blood supply and is used to keep the eggs and chicks warm. The male has no brood patch – incubation is the female's job and it begins when one egg has been laid and goes on for between 29 and 34 days. From time to time males do try to incubate eggs or brood young but not for long and they make no real contribution to the process. The male's job is to provide food for their incubating partner. The female, on the other hand, is a diligent incubator, turning the eggs from time to time and typically only leaving them for 5–10 minutes just two or three times a day for toileting, a preen, a bit of exercise or to retrieve food deposited by the male. Infertile eggs will be incubated for up to 60 days and even a close encounter with a predator is unlikely to force the female off her clutch.

As incubation nears completion, the young owlets can be heard from within the egg, and the female twitters quietly. The eggs hatch at different times, usually with gaps that mirror the laying pattern, though the intervals can be as long as two weeks. It takes 12–36 hours for a chick to extricate itself from its shell, ably assisted by its mother. She deals with the empty eggs by taking them out of the nest, eating them, moving them to one side or trampling them into the nest debris. European studies have recorded hatching rates of 70–87 per cent, while, in Utah, Marti recorded a lower rate of 63 per cent.

Looking after the young

A freshly hatched Barn Owl is pink, still has the egg tooth on its bill and has closed eyes. At this stage White-breasted Barn Owls weigh just 12 or 13g (0.4oz). Dark-breasted Barn Owls are a little heavier at around 14g

About three weeks after hatching, chicks develop thick downy plumage that may help to keep them warm after the mother stops brooding them. They will not start to look like an adult until they are about two months old.

When spots are good

It makes sense for male Dark-breasted Barn Owls to choose spotty females. The females can inherit their spottiness and Roulin *et al.* (2003) have discovered that the spottier the female, the better able the chicks are to cope with parasites. There are fewer blood-sucking flies in a spottier female's nest and the flies that are there produce fewer young.

(0.5oz), while North American Barn Owls average 16.4g (0.6oz). When they have dried out, a partial covering of off-white downy feathers becomes apparent, but there is not much of it and the female broods the youngsters to keep them warm for about 25 days. The owlets hatch at intervals, of course, so while the oldest owlet is over three weeks old at this point, the youngest might only have been out of the egg for 11 days. About this time, however, the second downy plumage begins to appear, and by the time the chick is around three weeks old, it has a thick

Overleaf: A male bringing food to the family nest. One of Roulin's studies found that paler-coloured males raise fewer chicks per nest and bring them less food than darker-coloured males.

coverage of white or off-white down and it seems that the siblings keep each other warm.

At first the chicks are weak – they can chitter quietly but are not able to lift themselves. Their yolk sac helps to nourish them for a couple of days and the female feeds them once every hour or so. The male delivers food and the female rips it up and feeds little pieces to the chicks by standing astride them with the food in her bill, triggering a grab and gulp response when the meat brushes the bristles above a hungry owlet's bill. For the first ten days or so, some females eat the chicks' faeces, but this may not always occur. Before long, the sounds of the nest have changed and 'beginner snores' are heard when the owlets hear the female's food-offering call. In these early stages the male may deliver more prey than is needed and food can build up at the nest. Colin Shawyer suspects that the male actively selects smaller prey items, such as Pygmy Shrews (*Sorex minutus*), from a cache to feed to the young when they are small.

Nest-site vocabulary diversifies in week two. Clicks and hisses are

The mother owl stands astride her chicks to feed them small morsels directly from her bill. By about two weeks old, the young can swallow whole animals. When the youngest chick can feed itself and does not need to be brooded both adults deliver food and head out to hunt again.

added to the repertoire and the owlets become very accomplished snorers! Before the week is out, the chicks are opening their eyes intermittently. By the time they are about two weeks old, they can stand and walk and progress from meals 'prepared' by the female to swallowing whole animals, something they may have been trying for a while.

When the youngest chick can feed itself and no longer needs brooding, the female normally leaves the nest. Now both parents can bring food in, dropping it off at the nest and then leaving. A Swiss study found that the male brought in 68 per cent of the meals, and Shawyer described the female's contribution as 'occasional'. Food can be delivered at an impressive rate and the parents have been known to deliver 33 animals in just 95 minutes, an average of one every 2.9 minutes. A more typical delivery rate, calculated from a number of studies, is an average of three animals per owlet per day.

Food deliveries in week three prompt enthusiastic snoring from the youngsters, though there may still be more food arriving than can be

eaten. The egg tooth is normally lost around this time and an owlet is now mobile enough to reverse to the side of the nest for toileting, if there is space to do so. By the end of a chick's third week, the quills of its primaries are obvious, and feather vanes can be seen at the tips of the quills when the bird is about three and a half weeks old. Owlets of this age begin to eat more food and defend themselves by lying on their back with their feet in the air and, as they get older, using their talons as weapons. By the time the chicks are about a month old, they are flapping their wings. Less food is brought to the nest and the youngsters' weight peaks when they are between five and six weeks old. At this point they are heavier than the adults, with a British owlet weighing as much as 400g (14.1oz), and are exploring outside the nest. Sleep is still an important part of the owlets' life and they sleep on their bellies! Usually their weight drops before they leave the nest but, given enough food, some will still be heavier than their parents when they fledge. New feathers grow, and when they are about two months old, the youngsters look much like the adults.

Chick development from newly hatched to nearly fledged.

Talking about food

Relationships between Barn Owl chicks are not just about cannibalism! There can be positive interactions too – chicks will preen each other and the older owlets have been known to feed their nest mates. Roulin and others have done a lot of work on the feeding behaviour of Barn Owl nestlings. Food can be hotly contested and they have found that the noisiest chick gets more than their fair share. The owlets will 'talk about food' when the adults are not at the nest, to decide who gets to eat the next delivery. When there are just two owlets in a nest, the hungriest one makes a lot of noise to show that it wants the next piece of food that is delivered. The other one keeps quiet and the noisy sibling gets the food. But then the unfed chick makes a lot of noise to show that it wants to eat the next delivery. Well-fed owlets are also known to reserve food for themselves by sitting on it! When food has been given to one chick, others may still want it. Research by Roulin and others published in 2008 found that the owlet with the food could be very noisy, especially if it was one of the younger birds, presumably to make it clear that it did not want the others to have any of the food. If the other chicks were particularly noisy, probably signalling their enthusiasm to get at the food, the one with the food ate it faster.

Leaving the nest and dispersal

In Europe, young Barn Owls take to the air for the first time when they are about eight weeks old, though the average fledging period of North American Barn Owls in Utah is about nine weeks. Their maiden flights are brief and inelegant but their flying skills quickly improve. They do not disperse immediately and the nest remains their base for around 3–6 more weeks and some roost nearby for even longer. They do, however, spend plenty of time out of the nest, pursuing siblings and screaming at unwelcome visitors, whether cat, fox or human.

Hunting is not entirely instinctive. The female takes on the teaching task, dropping food from a height, which the airborne young attempt to grab before it hits the ground. They fail and land on the ground to look for the food or try to find it from a lower hover. The youngsters may continue to be fed by the adults for a while but are usually able to look

An armful of Dutch Dark-breasted Barn Owl chicks.

after themselves by the time they are around 10–13 weeks old.

Not all of the young that hatch will make it to fledging. Studies in Europe have found survival rates that range from 65 per cent to 91 per cent and Marti's work in Utah produced a figure of 87 per cent.

Shawyer found that when his British Barn Owls finally dispersed, some did not go very far at first and were roosting close to the nest, and some were more adventurous and travelled greater distances straight away. Ringing (banding) data from young British Barn Owls suggest that it can be some months before their dispersal movements are complete. In North America, though, November can see some young Barn Owls settled at roosts that become their nest site.

A fledgling exploring a barn. After a few clumsy flights the fledglings become more competent fliers and will spend a lot of time outside the nest.

7 | Life and death

Most Barn Owls do not live very long. In Britain only 37 per cent survive their first year, though the odds improve significantly after that, when the annual survival rate increases to 72 per cent. Typically, if a bird makes it through its first year it will then live for another three. Figures from Scotland, the Netherlands, France and Germany tell a similar story, with only 15–35 per cent seeing their first birthday. Survival rates then increase year by year until a drop in the fifth year, when 45–65 per cent of the birds that have made it that far survive. North American data give an average life expectancy of about 21 months but this does include first-year mortality. Some Barn Owls live to grand old ages, with the record for a wild bird standing at 34 years.

What kills them?

Traffic kills a lot of Barn Owls – on roads and on railways. Roadside verges and their railway equivalent can be good foraging habitat but hunting there is a risky business. By the end of 1997, over 3,000 ringed (banded) Barn Owls had been recovered in Britain and Ireland. A cause of death could be ascribed to two-thirds of these and 82 per cent of them are thought to have been traffic casualties. French data showed that, of 1,598 road-killed birds, over 670 of them were Barn Owls, and in both the Netherlands and Hawaii, cars are believed to be the main Barn Owl killer.

Pesticides have also proved deadly. In decades past, organochlorine insecticides are known to have killed Barn Owls in the UK and the United States, for example, and could have contributed to a drop in English Barn Owl numbers in the 1950s and 1960s. In the 1980s and 1990s Barn Owls in New York State succumbed to organochlorines, and of 627 UK Barn Owls that died between 1963 and 1989, 56 are thought to have been killed by these infamous pesticides. Their impact on Barn Owl populations may have been small, however, compared to their impact on other birds of prey. In the 1970s a different chemical threat made its debut when a new generation of rodenticides was

Previous page: A classic view of a White-breasted Barn Owl at dusk.

launched to replace warfarin. Difenacoum was the first of these
chemicals and it is much more potent than warfarin. Bromadiolone,
brodifacoum and flucomafen became available later and these are more
potent than difenacoum – just three brodifacoum-contaminated mice
could be a lethal dose for a Barn Owl. The threat that
these chemicals pose to Barn Owls is unclear, though
they can, and have, killed Barn Owls. Studies in New
Jersey, though, where brodifacoum was being used, and
England, where brodifacoum, flucomafen and
difenacoum were put out, concluded that rodenticides
were not causing Barn Owls any problems. Coughing
up pellets may help – Newton found that 27 per cent of
the flucomafen eaten by a Barn Owl came back out in
the pellet. Data from the UK's Wildlife Incident
Investigation Scheme for 2009 includes only four Barn
Owl incidents and none of them blamed rodenticides,
though non-lethal amounts were found in several birds.
Molluscicides can be lethal to Barn Owls, though
Shawyer's study of Barn Owl mortality in Britain between 1982 and
1986 found that they killed many fewer birds than rodenticides.

Road accidents are a common
cause of death for Barn Owls
and many are also killed by
railways. The grassy verges
typical of these areas are
tempting, but dangerous,
hunting ground.

Despite being a very capable predator themselves, Barn Owls are
not always at the top of the food chain. In Europe Northern Goshawks
(*Accipiter gentilis*) and Eurasian Eagle-Owls (*Bubo bubo*) will take Barn
Owls, and Great Horned Owls (*B. virginianus*) will in North America.
Common Buzzards (*Buteo buteo*) and Peregrine Falcons (*Falco
peregrinus*) are known to have despatched them, and Colin Shawyer
reports a case of one meeting its end courtesy of a female Eurasian
Sparrowhawk (*Accipiter nisus*). Tawny Owls (*Strix aluco*) have also
killed them, and it has been said that this occurs when both species
want the same nest site. Mammals also need to be watched out for. In
North America, Northern Raccoons (*Procyon lotor*) will raid Barn Owl
nests, as will Beech Martens (*Martes foina*) in Europe. Colin Shawyer
listed Stoat (*Mustela erminea*), Weasel (*M. nivalis*) and American Mink

(*M. vison*) as nest raiders, plus, in an ironic twist of fate, Brown Rat (*Rattus norvegicus*). Perhaps surprisingly, domestic cats seem to have very little impact on Barn Owls.

In Shawyer's study almost one in four of the dead Barn Owls had met its end for no clear reason. Presumably these birds had died 'naturally' or had starved to death. A lot of them were sodden and may have been inexperienced youngsters with limited hunting skills.

Continuing the watery theme, some Barn Owls die while bathing – in cattle troughs and water butts, for example. Shawyer suggested that this is likely to be females cleaning themselves up after a couple of months spent incubating and brooding. Peter Wilkinson's assumption is simply that the Barn Owls are not able to judge the depth of the water and drown as a consequence. Others die after being trapped in buildings and some are casualties of collisions with wire fencing and overhead cables, or are electrocuted by power cables, particularly along railways.

In some parts of their range hard winters reduce Barn Owl numbers. In Britain, the winter of 2009–10 was particularly severe, and Kevin O'Hara, a Conservation Officer with the Northumberland Wildlife Trust (in the north of England), tells of over 50 dead Barn Owls that were picked up by members of the public or found dead at nest-box roost sites. Snow can make it difficult for Barn Owls to catch small mammals but in this case it was ice that brought about the owls' starvation, making it impossible for them to get at the voles underneath. Thankfully, unless there is a long series of severe winters, the Barn Owl population should be able to recover. Several studies have shown that a sudden drop in vole numbers has more impact on Barn Owl populations than a severe winter. In many parts of Europe, Barn Owl numbers vary from one year to the next, often mirroring the natural vole cycle. Taylor's work in Scotland revealed that Barn Owl and vole populations went up and down together, hitting a maximum once

Opposite: Inclement weather can be hazardous. In one study into Barn Owl mortality, it was not possible to establish a clear reason for death, but many of the casualies were sodden, suggesting that this may have been a factor in the owls' demise.

Drowning accounts for a number of Barn Owl deaths, but it is not entirely clear why this might be. Some authors suggest that most of these individuals are females trying to clean themselves after time brooding in the nest; an alternative explanation is that the birds misjudge the depth of the water and fall in.

every third year. A connection between Field Vole numbers and Barn Owl numbers has also been found in the Czech Republic and Germany.

Historically, Barn Owls have suffered persecution, be it pole trapping and shooting in late 19th-century Britain or shooting and destruction at the nest in 20th-century Spain. The problems they face currently are not totally understood, but collisions with traffic, habitat loss and the loss of suitable places to roost and nest need to be addressed to secure a positive future for this wonderful predator.

A helping hand?

Releasing captive-bred Barn Owls to augment a dwindling wild population has been fruitful to a limited degree but certainly not universally. It was tried in the United States, in the Midwest, but the vast majority of the 1,200 or so liberated birds were never seen again. Trudy Dockerty found that in Hertfordshire, England, between 1988 and 1992, the number of Barn Owls breeding did not change, even though 195 captive-bred birds had been released and they had raised 69 young. A study of English reintroductions over 21 years found that Barn Owl numbers had gone up but no one could be sure that the captive-bred birds had contributed to the change in fortune. Aside from the effectiveness of releasing captive-bred birds, there are other concerns. Released birds could bring non-native genetic material into an area – genes from Dark-breasted Barn Owls or the African subspecies could find their way into the British population, for example – or introduce diseases.

A helping hand

There are things that can be done to help Barn Owls. They need places to hunt, roost and nest. Areas of good hunting habitat need to be maintained, created and, as much as possible, joined up by corridors of rough grassland. During severe winter weather, when hunting can be very difficult for Barn Owls, piles of grain left at the

Snow can make hunting difficult and icy weather can cause severe problems if it is impossible to get at the mammals below.

edges of fields will feed rodents, which will feed hungry Barn Owls,
taking the place of farmyard stacks and grain stores, which were once
much more common and helped Barn Owls through tough winters.

A variety of methods have been suggested to improve road safety
for Barn Owls. One approach is to make sure that they fly higher
than the traffic. This can be done by judicious tree or hedge planting
or, at the construction stage, by ensuring that adjacent land is

sufficiently higher than the road. Removing plants from vole-rich verges is another option.

Barn Owls take to nest boxes readily and a well-planned nest-box programme can be very beneficial, particularly when there are few other nesting options and, of course, enough prey. There are many different designs (an example is shown opposite) and boxes can be installed in buildings, on the outside of them, in trees and on the top of poles. Nest sites can also be provided in the roof space of farm buildings that are being converted into houses. When they work well, nest boxes increase the number of Barn Owls in an area. Dutch farmers clearly cherish their Barn Owls – they feed them to help them survive severe winter weather and provide 'owl doors' so that they can get in and out of farm buildings.

You might not be able to get actively involved in providing nest boxes yourself but there may be a conservation organization in your area that is already involved in Barn Owl nest box work, which you could support. In the UK, the Hawk and Owl Trust has an 'Adopt a Box' scheme, which enables donors to do just that, or, if you would like advice about helping Barn Owls, contact the Barn Owl Conservation Network (BOCN), whose advisers will willingingly point you in the right direction.

Opposite: Barn Owls are often happy to adopt purpose-built boxes, especially if other nesting options are scarce.

Resources

Books

Beaman, M. and S. Madge, *The Handbook of Bird Identification for Europe and the Western Palearctic*, Christopher Helm, 1998.

Brown, A. and P. Grice, *Birds in England*, T & A D Poyser, 2005.

Bunn, D.S., A.B. Warburton and R.D.S. Wilson, *The Barn Owl*, T & A D Poyser, 1982.

del Hoyo, J., A. Elliott and J. Sargatal (eds), *Handbook of the Birds of the World*, Vol. 1, Ostrich to Ducks, Lynx Edicions, 1992.

del Hoyo, J., A. Elliott and J. Sargatal (eds), *Handbook of the Birds of the World*, Vol. 5, Barn-owls to Hummingbirds, Lynx Edicions, 1999.

Dewar, S.M. and C.R. Shawyer, *Boxes, Baskets and Platforms: Artificial nest sites for owls and other birds of prey*, The Hawk and Owl Trust, 2001.

Holden, P. and T. Cleeves, RSPB *Handbook of British Birds*, Christopher Helm, 2002.

Holloway, S. (compiler), *The Historical Atlas of Breeding Birds in Britain and Ireland: 1875–1900*, T & A D Poyser, 1996.

MacDonald, D. and P. Barrett, *Field Guide, Mammals*, HarperCollins, 1993.

Martin, J., *Barn Owls in Britain*, Whittet Books, 2008.

Mead, C., Owls, Whittet Books, 1987.

Mullarney, K., L. Svensson, D. Zetterstrom and P.J. Grant, *Bird Guide*, HarperCollins, 2000.

Perrins, C. (ed.), *The New Encyclopedia of Birds*, Oxford University Press, 2003.

Read, M. and J. Allsop, *The Barn Owl*, Blandford, 1995.

Shawyer, C., *The Barn Owl*, Arlequin Press, 1998.

Sinclair, I. and O. Langrand, *Birds of the Indian Ocean Islands*, 2003.

Slater, P., P. Slater and R. Slater, *The Slater Field Guide to Australian Birds*, Reed New Holland, 2007.

Sparks, J. and T. Soper, *Owls, Their Natural and Unnatural History*, David & Charles, 1970.

Taylor, I., *Barn Owls: predator–prey relationships and conservation*, Cambridge University Press, 1994.

Tucker, G.M. and M.F. Heath with L. Tomialoj and R.F.A. Grimmett, *Birds in Europe: their Conservation Status*, BirdLife International, 1994.

Scientific papers

Millsap, B.A. and P.A. Millsap, 'Burrow nesting by Common Barn-Owls in north central Colorado', *The Condor*, Vol. 89, No. 3, Aug. 1987, 668–70. Viewed at www.jstor.org/stable/1368658.

Roulin, A., C. Colliard, F. Russier, M. Fleury and V. Grandjean, 'Sib–sib communication and the risk of prey theft in the barn owl Tyto alba', Journal of Avian Biology, Vol. 39, 2008, 593–8.

Newsletters

The Barn Owl Trust, *Feedback*, Issue number 38, Autumn 2007.

DVD-ROMS

Birds of the Western Palearctic Interactive, version 2.0, BirdGuides, 2006.

Online

Clements, J.F., T.S. Schulenberg, M.J. Iliff, B.L. Sullivan and C.L. Wood, *The Clements Checklist of Birds of the World*: Version 6.4. 2009. Downloaded on 22 April 2010 from www.birds.cornell.edu/clementschecklist/Clements%206.4.xls/view

Gill, F. and D. Donsker (eds), IOC *World Bird Names* (version 2.4). 2010. Available at www.worldbirdnames.org [Accessed 31 March 2010].

Marti, C.D., A.F. Poole and L.R. Bevier, *Barn Owl* (Tyto alba), *The Birds of North America Online* (A. Poole, Ed.), 2005. Cornell Lab of Ornithology. Retrieved from *The Birds of North America Online*: http://bna.birds.cornell.edu/bna/species/001 doi:10.2173/bna.1

Barn Owls and Kestrels are birds of a feather, on: www.metro.co.uk

Barn Owl Conservation Network: www.bocn.org

BirdLife International: www.birdlife.org

British Trust for Ornithology: www.bto.org

The Cornell Lab of Ornithology, All About Birds: www.allaboutbirds.org

Hawk and Owl Trust: www.hawkandowl.org

Smithsonian Institution, National Museum of Natural History, North American Mammals: www.mnh.si.edu/mna/

Wildlife Incident Investigation Scheme: www.pesticides.gov.uk/environment.asp?id=58

Index